H

The Field of Geography

General Editors: W. B. MORGAN
and J. C. PUGH

Cartographic Methods

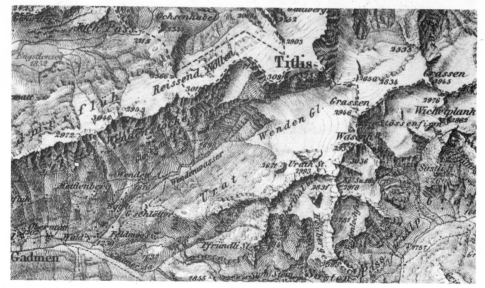

Extracts from two official topographic maps of the same region of Switzerland.
(above) 1864 1/100,000 map, sheet 13, relief is shown principally by hachuring. This series was compiled under the direction of G. H. Dufour and the maps are frequently referred to as 'Dufour maps'.
(below) The same area on a recent map (1967 1/25,000 map, sheet 1211, Meiental) in which contouring, hill shading and rock-drawing are combined to give an excellent impression of the topography of the region.

Cartographic Methods

G. R. P. LAWRENCE

METHUEN & CO LTD

First published by Methuen & Co Ltd,
11 New Fetter Lane, London EC4 1971
© 1971 G. R. P. Lawrence
Printed in Great Britain by
Butler & Tanner Ltd, Frome and London

SBN (Hardback)   416 07100 7
SBN (Paperback)   416 07110 4

Contents

The Field of Geography

Progress in modern geography has brought rapid changes in course work. At the same time the considerable increase in students at colleges and universities has brought a heavy and sometimes intolerable demand on library resources. The need for cheap textbooks introducing techniques, concepts and principles in the many divisions of the subject is growing and is likely to continue to do so. Much post-school teaching is hierarchical, treating the subject at progressively more specialised levels. This series provides textbooks to serve the hierarchy and to provide therefore for a variety of needs. In consequence some of the books may appear to overlap, treating in part of similar principles or problems, but at different levels of generalisation. However, it is not our attention to produce a series of exclusive works, the collection of which will provide the reader with a 'complete geography', but rather to serve the needs of today's geography students who mostly require some common general basis together with a selection of specialised studies.

Between the 'old' and the 'new' geographies there is no clear division. There is instead a wide spectrum of ideas and opinions concerning the development of teaching in geography. We hope to show something of that spectrum in the series, but necessarily its existence must create differences of treatment as between authors. There is no general series view or theme. Each book is the product of its author's opinions and must stand on its own merits.

W. B. MORGAN

J. C. PUGH

University of London,
King's College
August 1971

Text figures, tables and map extracts

Preface

This book sets out to provide a general survey of those aspects of cartography of relevance and importance to the geographer. Just as the study of geography embraces a variety of different outlooks, cartography involves its practitioners in a number of subjects, ranging from the intensely practical work of the map-printer to the artistic demands of map design. The geographer, in handling and making maps for his own purposes of analysis and also of data presentation, provides a further meaning to the term cartographer. The entire field of cartography is therefore large and a complete handbook would run into several volumes. I have therefore tried to select a number of pertinent aspects in a short space, and to provide a few signposts to the more detailed treatment of the subject, particularly by quoting some of the more accessible references in other books and papers.

Although some of the illustrations in this book are taken from published material and are appropriately acknowledged, the great majority of the diagrams have been drawn by my wife to whom I am greatly indebted, not only for the draughting but also for much help and encouragement in preparing this book. Thanks are also due to Miss Pat Aylott, departmental photographer, Geography department, University of London King's College for help in preparing photographic copies. Much other assistance was also provided by the cartographers, technical and secretarial staff in the same department and is also gratefully acknowledged. Finally, I am greatly indebted to both Professor J. C. Pugh and Dr. W. B. Morgan as well as the publishers for their editorial advice and assistance.

1971

King's College
London

Acknowledgments

The author and publisher would like to thank the following for permission to reproduce copyright material:

Professor E. H. Brown for fig. 1. 4d. Dr M. C. Storrie and Dr C. I. Jackson for fig. 3.4. Frontispiece reproduced by permission of the Topographical Survey of Switzerland. The Controller of H. M. Stationery Office (Crown Copyright reserved) for figs. 2.1, 2.11, 3.6a, 3.7, 4.4, 9.3 and material used in App. I. Pergamon Press Ltd for fig 3.2. U.S. Geological Survey for fig. 3.6b and material used in Appendix III. Topografische Dienst, Delft, for fig. 4.1. Landesvermessungsamt Rheinland-Pfalz for fig. 4.2, and material used in App. II. University of London Press for fig. 7.1. The Institute of British Geographers for figs. 9.9 and 10.7. Geographical Publications Ltd for fig. 10.5. George Philip and Son Ltd for fig. 11.3. Letraset Ltd for figs. 12.6a and b. Rexel Ltd for fig. 12.10. Institut Géographique National for material used in App. II.

Introduction

The geographer's field of study is sometimes precisely, sometimes loosely defined. The subject matter involved deals not only with the surface features of the earth but also with the interrelationships between man and his environment. The presentation of these facts and their analysis in depth often involve specialisation in disciplines beyond the realm of geography, but the geographer is usually recognised as the expert when spatial patterns and relationships are under discussion. Although these patterns may be represented in the abstract way made possible by mathematical and statistical studies it is often simpler to perceive them in a pictorial manner. Such is the field of mapping to the geographer – both as a source of information and as a method of presentation. Even so, the whole range of cartography covers a wide field. It is convenient to divide this field into those sections which deal with the presentation of information and those which consider methods of studying mapped data. Obviously the former is coupled with a study of one of the basic sources of geographical data – the topographic map – and here one enters the field of surveying as well as cartography. Techniques of analysis depend to a certain extent on the interest within the subject – thus the geomorphologist's approach may well differ from that of the historical or the urban geographer. Nevertheless there are many points of similarity in these analyses as all are concerned with distributions in geographical space.

The first section of this book considers the basic function of any map or cartographic system – the need for scale representation and the presentation of information according to cartographic rules and conventions. How these rules are applied in various published maps is considered in the second section, which examines British and overseas map series, as well as atlases. Cartographic methods of study involving some modern quantitative techniques make up the third section and the concluding section briefly discusses a number of ways in which data may be presented in cartographic form for research or publication.

Part I
General characteristics

SCALES

GRIDS

REPRESENTATION OF RELIEF

ACTIVITIES OF MAN

Fig. 1.1. Vertical air photograph of part of the inner urban area of Christchurch, New Zealand, together with monochrome enlargement from the 1/63,360 topographic map of the same area. This area is very nearly level so height distortions due to local relief are negligible. The greater detail of the black and white air photograph can be easily appreciated. The scale of the air photograph is approximately 1/20,000 and the map 1/50,000.

1 The basic framework – scales and grids

A map is normally defined as a representation to scale of the features of the surface of the earth. Such features present a number of problems to the map-maker since they include not only visible and tangible items such as hills, valleys, buildings and roads, but also important yet invisible features such as boundaries and frontiers. Many of the visible features are well portrayed in a photographic view. Such a view, almost to scale, is given by an aerial photograph taken from a point directly above, i.e. a vertical aerial photograph. If we examine the topographic map of the same area as the photograph a number of similarities and differences may be noted (fig. 1.1).

The aerial photograph shows all the visible details of the land surface; presenting the information to the user by a series of *tones* – either shades of grey, white and black or of colours. In comparison, much map data is presented in the form of information concerning a particular *point* – shown with a symbol – or concerning a *line* – represented by a printed line on the map. The map can also show arbitrary information which does not exist on the ground, place-names, boundaries and the like. It is, however, highly selective in character and shows only that information which the map-maker desires to present to the map-user. On the other hand, a single aerial photograph is inadequate to portray the land relief and other height variations which play an important part in every landscape. This can be overcome by using stereoscopic pairs of aerial photographs from which the user can obtain a three-dimensional mental picture or space model. Again, the aerial photograph shows all the details of the land surface recorded by the camera. In many cases the resulting picture is so complex as to make the identification of features a difficult procedure.

Both the vertical aerial photograph and the map are scale representations of the earth's surface. By this is meant that accurate measurements made on the map are proportional to their true values on the surface of the ground, or that every point is correctly positioned relative to all other points. Measurements from aerial photographs are of lesser accuracy than those on maps because of various distortions; but nearly all maps are prepared as scale representations on a plane surface,

3

normally the surface representing sea-level. On small scale maps covering extensive areas of the earth's surface, where the problem of curvature of that surface arises, it is necessary to consider the map projection which has been used and the properties this possesses. The topics of mapping from ground surveys and from aerial photographs as well as map projections are studies in themselves. Therefore, for the purposes of the discussions in this volume it will be assumed that the distortions resulting from the curved surfaces of the earth have been removed by the map-maker's choice of projection or that the scale of the map is sufficiently great for these errors to be disregarded.

Representation of scale

Scale is defined as the relationship between a distance measured on the map and the true distance on the ground. The natural scale is usually expressed as a ratio or 'Representative Fraction' and metric relationships are increasingly used, e.g. 1/10,000 or 1/50,000. Nevertheless, many existing maps in the English-speaking world include Imperial units, e.g. 'one inch to the mile' and these give rise to such ratios as 1/63,360, there being 63,360 inches in one mile. Other unusual scales may be encountered in old maps or maps prepared to be printed on a given sheet size, such as older nautical charts or atlas maps. Divergences from the intended natural scale may occur in the production of aerial photographs by reason of problems of maintaining flying height or large differences in land height. If no scale value is given for a map,

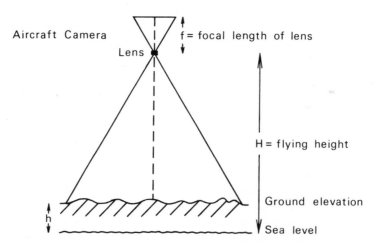

Fig. 1.2. Relationship between aircraft flying height, focal length of camera lens and scale of the air photograph.

either by a fraction, printed scale bar or statement in words, it must be calculated by comparing a measured length on the map with its known length on the ground. In the case of aerial photographs the flying height (H) and the focal length (f) of the camera used may be quoted in the marginal information. The scale can then be calculated from the expression $\dfrac{f}{H-h}$ where h is the height of the ground above mean sea level (fig. 1.2). Care must be taken in all these calculations to ensure that comparable units are used (inches, feet, miles, or metres, kilometres, etc.) e.g.,

(a) 1 cm on the map represents 1 km on the ground gives a Representative Fraction of 1/100,000.

(b) Aerial photograph taken from a height of 5,000 ft by a camera of focal length 6 ins will have a scale of 1/10,000 if the area photographed is at sea level, or

$$\frac{6 \text{ ins}}{(5,000 \text{ ft}) - (1,000 \text{ ft})} = \frac{1}{8,000}$$

if the area photographed is 1,000 feet above sea level.

(c) If 0·5 ins on the map represents 2 miles on the ground, the scale is calculated as 0·5 represents 2 × 63,360 ins or 1/253,440, approximately 1/250,000 (Quarter Million).

Map scales are extremely varied. In general, maps at a very large scale, e.g. 1/500 or 1/1,250 are referred to as *plans* and show detail precisely and with no exaggeration and little variation from the ground shapes they portray. The map projection used by such plans will be barely obvious to the user. Small scale maps such as those of countries or whole continents will require to be presented on a map projection graticule and are best referred to as *atlas* maps. It is generally found that simplifications of ground detail begins between the scales of 1/5,000 and 1/10,000 and once it has been found necessary to introduce generalisations of shape or to omit items of detail the cartographer's work becomes subjective. The degree of subjectivity may be reduced by a form of shorthand in which particular objects or features are replaced by symbols. Frequently, however, it is necessary to suppress information in order that chosen symbols, or even the necessary description, naming or lettering, may be placed on the map.

In connection with map scale the words 'small', 'medium' and 'large' are frequently used. There is little general agreement as to where the dividing lines come. Recent usage by British organisations has been as shown in the table below:

	Ordnance Survey	Directorate of Military Surveys	Directorate of Overseas Surveys
Large scale	1/10,560	1/75,000	1/25,000
Medium scale	—	1/75,000 to 1/600,000	1/25,000 to 1/125,000
Small scale	1/25,000	1/600,000	1/125,000

It may sometimes be helpful to recognise a fourth category of 'atlas scale' for maps at much smaller scales, e.g. 1/1,000,000 or smaller.

This discussion of the significance of the scale of a map must also emphasise the fact that any line or symbol drawn on a map has a definite size. Even the finest printed line will represent a strip of land and thus reduce the amount of space available on the map for other information. For example, a line 1/100 in wide on the map drawn at the scale of one inch to the mile (1/63,360) will represent a band more than 50 ft wide. Or, even a line 0·1 mm wide will cover a width of 5 m on a 1/50,000 map. The aerial photograph shows many lines by the boundary between different tones or colours and often in these cases no measurable thickness can be resolved. On the printed map, however, the boundaries between land parcels, road edges, etc. are shown as printed lines. It is customary to regard these lines as having no thickness but errors may arise, for example, in measuring distributions and areas on such maps.

The problem of scale is encountered at some stage in all geographical studies. One possible framework which can link all variations in size from the entire earth's surface down to the individual land parcel is the *G-scale* (Haggett 1965). Using this scale and a map format of 1,000 sq cm in area the following table indicates variations in size of land surface covered by maps at different scales. Thus a 1/10,000 map of this size covers 10 sq km whilst a 1/100,000 map will include 1,000 sq km. The respective G-scale values for these two cases are 7·7074 and 5·7074.

G scale value for map sheet of 1,000 sq cms format

Large scale plans (e.g. 1/2,500)	G scale 8·5033
Medium scale maps (e.g. 1/25,000)	G scale 6·5033
Small scale maps (e.g. 1/250,000)	G scale 4·5033
Atlas maps (e.g. 1/2,500,000)	G scale 2·5033
(see also fig. 1.3).	

Maps are of very varied size and appearance as well as of differing scales, but in general it is possible to distinguish between two major

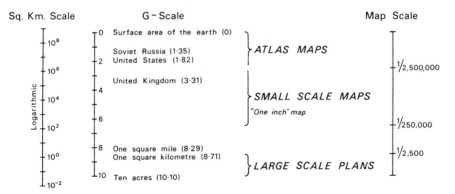

Fig. 1.3. Geographical scale, map scale and area. Diagram based on the concept of 'G-scale' (Haggett 1965).

categories of maps. On the one hand there are maps which depict as much of the surface features of a particular area as is possible within the limitations of the scale. Such maps are called *topographic* maps and familiar examples of these are the 1/25,000 and 1/50,000 national map series produced by many countries, and the one inch to the mile Ordnance Survey map of Great Britain.

On the other hand many maps are designed for specialised purposes. They may be drawn to indicate certain distributions and/or to supplement the basic data of the topographic map. In many cases all or part of the topography will be shown as a base map with the additional information drawn or overprinted upon this framework. Examples of such cases are maps of geological outcrops, land utilisation, and soil and vegetation maps. Again, the particular purpose of the map may be best served with only the minimum framework of reference points. The common factor in all these maps is that they have a single theme of study and they are therefore grouped together as *thematic* maps.

It is clear from the above that the so-called topographic maps are themselves thematic, in that they have been prepared for a specific purpose. This purpose is to present an overall view of the topography of the particular area for a variety of users – and in many cases the needs of a particular user – e.g. the military authorities – may have been paramount. The picture of a countryside which they portray is limited in various ways. Although a policy of 'continuous revision' may be adopted, as with most current Ordnance Survey productions (see pp. 52-55), the topographic map presents a 'synoptic' picture of the land surface at the time of the survey. For some geographical purposes this may be adequate, e.g. a series of maps of different dates will be of value to any study of the historical development of a region. This

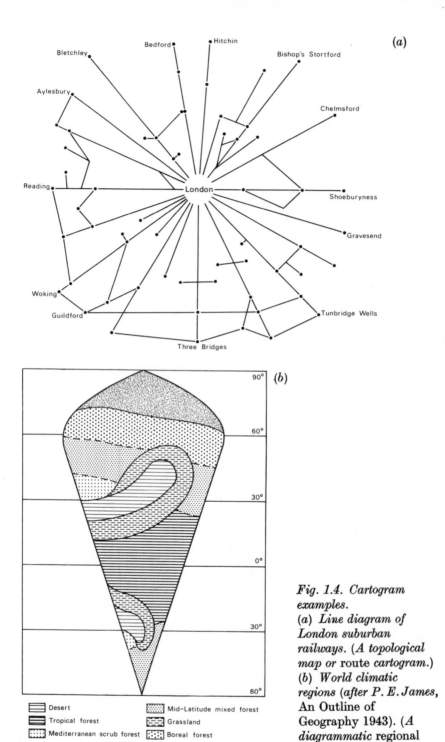

Fig. 1.4. Cartogram
examples.
(a) Line diagram of
London suburban
railways. (A topological
map or route cartogram.)
(b) World climatic
regions (after P. E. James,
An Outline of
Geography 1943). (A
diagrammatic regional
cartogram.)

(c)

- Minor Service Centres
- Principal Service Centres
→ Hierarchal Relations

Leeuwarden   Groningen

Zwolle

Amsterdam

Enschede

Leiden

The Hague   Utrecht

Rotterdam   Arnhem

Nijmegen

s Hertogenbosch

Breda

Middelburg   • Tilburg

Bergen-op-Zoom   Eindhoven

Heerlen

Maastricht

N

0   50 miles

0   50 kilometres

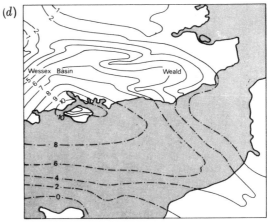

(d)

Wessex Basin   Weald

(c) *Urban hierarchies in the Netherlands. (Cartogram of a stage in urban analysis.)*
(d) *Contours of the Palaeozoic platform (after E. H. Brown in* Guide to London Excursions, I.G.U. 1964). (*A theoretical cartogram.*)

static framework of the topographic map is generally adequate as a basis for presenting the results of a geographical study but in many branches of modern geography an emphasis is placed on dynamic characteristics, such as changes, processes, development and functional relationships or the potential for these.

Board (1967, in Chorley & Haggett) distinguishes between *single-purpose* and *multi-purpose* maps. Although thematic maps can belong to the latter category they are more usually single-purpose in character whilst topographic maps are multi-purpose. For various study purposes geographers, economists, planners, etc. need to prepare and/or make use of thematic maps and it is useful to further subdivide this category. The *diagrammatic* presentation of information can be based on its real spatial relationships (i.e. utilising natural positions on the earth's surface) or according to relationships obtained from e.g. theoretical considerations, non-Euclidean geometries, model-building theories and the like. Such maps are frequently referred to as *cartograms* and they have also been called *topological* maps (Cole and King 1968). The selective presentation of results of study or of material for purposes of further study in diagram form might be separately classed as *analytic* maps. Examples of these categories are:

> *Cartograms* – British Rail, London Transport route maps
> *Diagrammatic* – Urban hierarchies – Functional regions
> *Analytic* – Generalised contours

In all mapping it is necessary to be able to locate individual points with some degree of precision and for the map-maker the precise location of detail will be one of the major stages in compilation from field data. The printed map will eventually be consulted by the map-user who will wish to extract information from it, or to locate further data on it. The plane surface of the map sheet will thus need a reference framework by which points can be sited or found. On the surface of the earth an arbitrary and well known system of references is used – lines of latitude and longitude, which form a graticule. These are also generally used on map series but they suffer from a major defect. The earth's surface is curved and although the lines of latitude and longitude are straight and intersect at right angles on this surface they cannot do so on a flat sheet and also preserve angular relationships desired by the map-maker. However, two dimensional reference systems used to locate points are normally based on a rectilinear grid and measurements in units of distance (kilometres, miles, etc.) to the nearest grid line can be made correctly and easily, whilst measurements of latitude or longitude to graticule lines require greater care as, for example, the lengths along the parallels decrease as one moves from the Equator to the Poles.

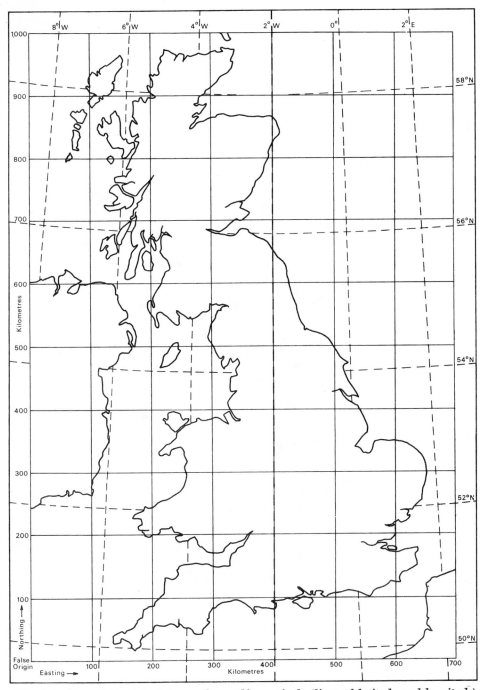

Fig. 1.5. The relationship between the earth's graticule (lines of latitude and longitude) and an arbitrary grid, in this case the National Grid of the Ordnance Survey of Great Britain. Only the southern 1,000 km of the grid are indicated; the northern continuation to include the Orkneys and Shetland is not shown. Note the coincidence of the 2°W meridian and the 400 km grid line, also the position of the false origin so that all references are positive. The true origin of the orthomorphic Transverse Mercator projection on which the grid is based is at 49°N and 2°W.

The choice of a map grid system for referencing is made by the map-maker and will be made after consideration of problems of map projection, layout of the sheets forming a map series, desirability or otherwise of linking with map series of adjacent countries, etc. The map grid best known to British users is the National Grid Reference System used by the Ordnance Survey.

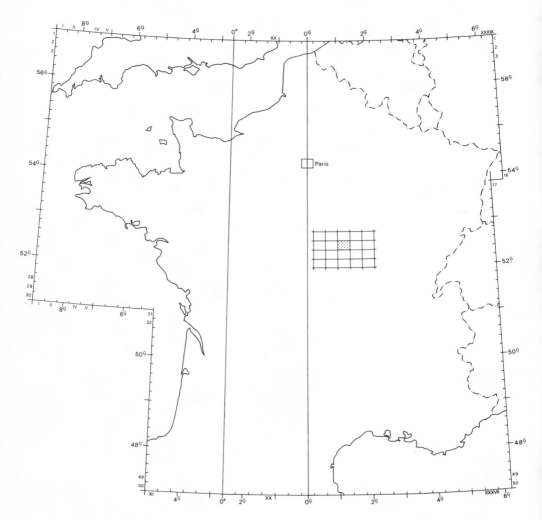

Fig. 1.6. The arrangement of the topographic map sheets of France (1/50,000 scale) and their relationship to the graticule. Note that the prime meridian is that of Paris and that latitude and longitude are measured in grads (100 grad = 90°). Map sheets are referred to by combining Roman numbers (columns) and Arabic (rows). Shaded map is thus XXVI–23.

Conventionally in this form of grid, referencing points are sited and referred to in a version of *Cartesian co-ordinates* or rectangular grid co-ordinates. Thus the extreme south-western point of the system is taken as the origin and all points are related to this origin. In actual fact this origin in the National Grid is a 'false origin' in order to remove problems of negative co-ordinates. The true origin is at 2°W long. and 49°N latitude (fig. 1.5). In order to reduce scale error at the margins of the grid the scale along the central meridian has been reduced by 0·04% as the scale increases outwards from the central meridian; consequently the modification ensures only small scale errors, negative or positive, instead of errors all positive in sign but of greater percentage error away from the central meridian. These details are adequately explained in the HMSO publication 'The projection for Ordnance Survey Maps and plans and the National Reference System', and need not be considered further here. The National Grid System of giving references at various scales is given in Appendix I. It should be noted that in all reference systems the distance *east* from the origin is quoted first, followed by the distance *north*, i.e. 'Eastings' before 'Northings'.

Other systems of locating points are, or have been used on various map series. These mostly fall into the category of locating points to a specific square which is found by a combination of identifying letters and numbers in the map margins. It is less convenient than that depending on Cartesian co-ordinates and is obviously less precise. A variation on this is used by some mapping agencies in order to refer to map sheets within the series. This is well exemplified by the French national map series (fig. 1.6) in which a combination of numbers in Roman and Arabic form are used. Those in Roman type give the north–south column, which is also linked with the meridians on a Conic projection graticule, whilst the west–east lines are shown by Arabic numerals.

The importance of co-ordinate reference location systems is increasingly marked with the use of data banks and computer methods of storing and analysing data. A rectangular system of co-ordinates is used in most cases and it may be necessary to distinguish between the co-ordinates used on the map and those used for computations. The necessary qualities of such a location referencing system are:

(i) The specified point must be unique.
(ii) Relative positions must be capable of identification.
(iii) The system must be capable of use with electronic data processing methods. (Maps produced by computer are considered on pp. 49–51.)

2 Representation of relief, drainage and coasts

All topographic information can be seen to relate to a specific point, line or area, and many cartographic methods apply to the depiction of both the natural and artificial features of the topography. Nevertheless, some methods of representation have been developed for particular features, e.g. land contouring is applied to a visible surface seen either in the field or in the space model of the stereo-viewer whilst other isolines represent less tangible shapes. Similarly, special symbols and mapping methods have been developed for representing hydrographic features.

This chapter will therefore consider in particular various methods of representing the natural features of the physical topography, i.e. the relief, drainage and coastlines. The representation of distributions, point and line information other than that relating to natural features will be considered in chapter 3.

Representation of relief

This is probably the most difficult problem in mapping (and map reading) and stems from the basic need to present the three dimensions of the landscape on the flat, two-dimensional paper form in which the map appears (fig. 2.1).

Methods used to overcome this difficulty can be considered under two main headings – those in which a visual *impression* of relief is presented and those in which precise *values* can be set out.

However, it is first necessary to consider briefly the possible data available to the map maker. Height information may be provided by precise survey methods of levelling in which the exact height of a series of points is determined. These points may be recorded on the ground as bench marks and this elevation is related to a datum or mean sea level at some specified point. In Great Britain this datum is a level 15·588 ft below a bench mark adjacent to the Tidal Observatory at Newlyn in Cornwall. This value was calculated by taking the mean value of hourly sea level values between 1915 and 1921. A previous datum was established at Liverpool and on old large scale Ordnance Survey maps discrepancies may arise between values calculated from

14

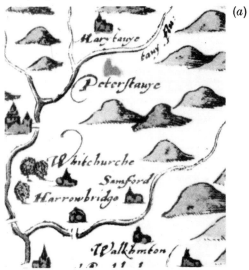

(a)

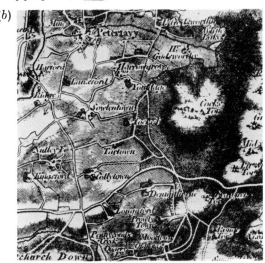

(b)

Fig. 2.1. Three map extracts to
show relief representation methods
over the centuries.
(a) Extract from Saxton's map of
Devon (1575): hill symbols.
(b) The same area on the first One
inch to the mile Ordnance Survey
map of Dartmoor (1809): hachures.
(c) Monochrome print of the same
area on the Dartmoor tourist
edition One inch to the mile map
(1967): spot heights, hill shading,
contours and layer shading.

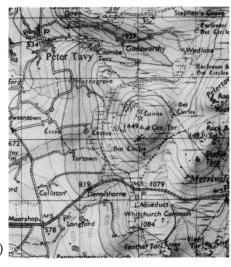

(c)

these two datum points. Corresponding datum values are established by all national mapping organisations and again discrepancies may be encountered when considering maps of adjacent countries, e.g. between the French and German topographical series in north-west Europe (fig. 6.1). Surveyed height information from ground surveys based on such data is therefore available at a large number of selected points: there is no information on the way in which height variations take place between these points.

Much mapping today is carried out from aerial photography. Stereoscopic examination of pairs of air photographs provides a visual impression of the landscape in miniature and height differences can be calculated by measuring parallax displacements. Once again fairly precise values for individual points can be obtained and the cartographer has the added advantage of a view of the three-dimensional scene in the overlapping pair of air photographs. He can interpret this 'space model' visually, to obtain positions of breaks of slope, etc., and can plot contour lines (see below) by steering along them the floating point in the plotting machine. More advanced methods of mapping from air photographs (e.g. using the orthoprojector) enable contours to be plotted semi-automatically (Petrie 1962 and 1966).

Quantitative representation of height

It is thus possible to show on a map the exact value of heights of various points on the ground so represented. There would be no visual impression of landforms in such a map. These values or precise *spot heights* are, however, extremely important, e.g. summit heights, depths of valleys. The conventional representation of relief stemming from such information is that of contouring: a contour line is one on which all points are of equal vertical distance above or below a datum. It is one of the most well known forms of *isogram*. Strictly speaking the term contour should be reserved for those lines which have been carefully surveyed on the ground as having the particular value with which they are labelled. Precise photogrammetric methods make it possible nowadays to draw contour lines of considerable accuracy, depending chiefly on the scale of the air photography involved.

It is also possible to interpolate lines of height from a close network of spot height values and computer programs can do this automatically. The normal assumption made in such interpolation is that height values change at a steady rate between neighbouring spot heights. The resulting isogram (or isoline) cannot therefore be given the same degree of precision and should be termed a *form line*. If the contour line had been obtained by precise measurement, as by levelling and pegging out in the field followed by detailed surveying, it then merits the term *isometric* line, i.e. a line on the map obtained by measurements at a series of points. Spot heights and surveyed contour lines thus give precise values of the height of the land surface and form quantitative data on the map. Contours are assumed to be accurate but errors can occur. If an error (dh) is made in height then an error is made in position (dl) which is related to the ground slope

$$\left(\text{since } \tan \alpha = \frac{dh}{dl}\right)$$

(fig. 2.2). Imhof (1965) has demonstrated a variety of possible contour errors arising from such cases as systematic errors in position or height, faulty contour intervals, accidental and gross errors. The majority of contours on present-day maps are compiled from stereoscopic air photographs and refinements in photography and photogrammetric techniques have led to a high degree of accuracy in their compilation.

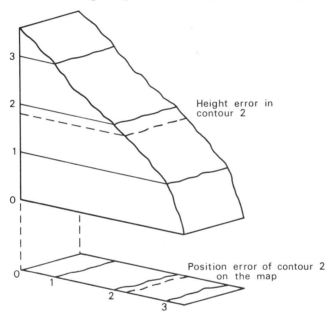

Fig. 2.2. Diagram to illustrate the effect on the plan position of a contour line following errors made in height determination (or vice versa).

Contour lines are normally shown on maps and plans at regulaɪ intervals; the interval between successive contours known as the vertical interval (V.I.) and indicated on maps produced in non-English languages as 'equidistance' (or its equivalent). The appropriate spacing varies according to the terrain and needs of the map-maker and map-user. In low-lying areas a frequent contour interval is clearly necessary whereas in hilly areas this could lead to confusion. Thus on a 1/25,000

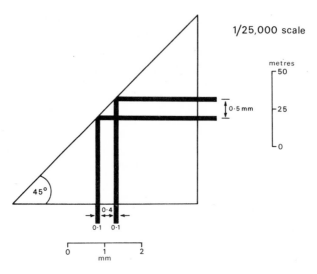

Fig. 2.3. Enlarged cross-section to show relationship between printed contour lines (base) and their true vertical interval, for slope of 45°.

map, if one allows for the printed line to be 0·1 mm wide and a spacing of 0·4 mm to be capable of being resolved by the eye between the centres of the lines representing two adjacent contours, then a contour interval of 12·5 m might be used even for slopes of up to 45° (fig. 2.3). An approximate formula which can be used for obtaining a possible contour interval is given below, assuming that a more realistic resolution distance is 1 mm.

$$\text{C.I.} = \frac{S \tan \theta}{2 \times 1,000} \text{ m}$$

where S is the denominator of the map Representative Fraction and θ the steepest slope to be represented.
e.g. map scale 1/50,000

$$\text{C.I.} = \frac{50,000}{2,000} \tan 45°$$

$$= 25 \text{ m}$$

Imhof (1957) has further shown that the ideal contour interval is in fact given by the formula C.I. $= n \log n \tan \theta$ m where

$$n = \sqrt{\frac{S + 1}{100}}$$

and S is the denominator of the map scale Representative Fraction. An angle of 45° for mountainous areas, 26° for areas of medium relief and 9° for flat relief has been used to calculate appropriate intervals (e.g. Alpine Europe, Hercynian Europe and European plain respectively) (fig. 2.4).

Scale	A					B	C
	1	2	3	4	5	6	7
1/2,000	1, 0	2	2, 7	2	1, 0	1, 0	0, 5
1/5,000	2, 5	5	5, 7	5	2, 5	2	1
1/10,000	5, 0	10	10	10	5	5	2
1/25,000	12, 5	10, 20	19	20	10	10	2, 5
1/50,000	25	20, 25	29	20	10	(10)	5
				30	15	20	
1/100,000	50	50	47	50	25	25	(5)
1/250,000	125	100	85	100	50	50	10
							(20)
1/1,000,000	500	200	200	200	100	100	20

Fig. 2.4. Table of contour spacing in metres for various map scales and for three groups of terrain (after Imhof 1965).
A. Mountainous regions (slopes up to 45°).
    1. Smallest contour interval for ease of drawing (see fig. 2.3).
    2. Most commonly used contour interval.
    3. Theoretical contour interval calculated from formula (p. 18).
    4. Recommended contour interval for principal contours.
    5. Recommended contour interval for intermediate contours where necessary to depict features otherwise omitted.
B. Areas of medium relief, suggested contour interval.
C. Areas of low relief, suggested contour interval.

The contour lines shown for a given area often do not adequately depict all the relief features; for example, locally important hollows or terrace features may not be represented at all because the selective contour interval omits them. This can be remedied by introducing additional 'in-between' contours at half or even one-quarter, half and three-quarters of the normal contour intervals. Such extra lines must, however, be in a different thickness or style of line and should only be used where necessary. It is also normal practice to label contour lines with their height value in such a way that the numbers stand with their base on the downhill side. Again, contour values may be placed on the line or, preferably, a gap made in the line. Thickening of certain contour lines helps in the interpretation of values but must be at a regular interval (fig. 2.5).

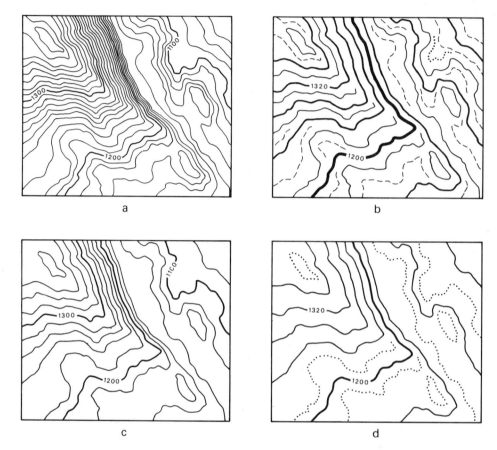

Fig. 2.5. Examples of different contour intervals depicting the same relief pattern.

(a) *Contours at 10 m interval, thickened at 100 m.*

(b) *Contours at 40 m interval, thickened at 200 m, and with intermediate local contours at 10 and 20 m spacings.*

(c) *Contours at 20 m interval, thickened at 100 m.*

(d) *Contours at 40 m interval, with intermediate local contours at 20 m spacing only.*

(a) *shows signs of coalescence of contours and hence at the scale of reproduction fig. (c) would be most suitable. At a smaller scale (b) or (d) would be more satisfactory, (d) showing the better generalisation. Note that the 40 m interval means that only certain 'hundred metre' contours will be present, i.e. in this case 1200 m only. (Based on Imhof.)*

Reliability of contours

The reliability by which contours have been mapped has increased with improved methods of surveying and of photogrammetry since the nineteenth century: with the result that the mean error of contouring on 1/25,000 maps has improved from $\pm(4 + 25 \tan \theta)$ before 1890 to $\pm(0\cdot5 + 5 \tan \theta)$ at the present time. (i.e. on slopes of 45° an accuracy then of ±29 m, now of $\pm5\cdot5$ m and on slopes of 10° an accuracy then of ±9 metres, now of $\pm1\cdot5$ m). The following table indicates the accuracy of contours on some common topographic map series.

Country	Scale	Mean error in metres
Germany	1/5,000	$\pm(0, 4 + 5 \tan \alpha)$
Switzerland	1/10,000	$\pm(1 + 3 \tan \alpha)$
Great Britain	1/10,560	$\pm \sqrt{1, 8^2 + 3, 0^2 \tan^2 \alpha}$
France	1/20,000	$\pm(0, 4 + 3 \tan \alpha)$
Germany	1/25,000	$\pm(0, 5 + 5 \tan \alpha)$
Great Britain	1/25,000	$\pm \sqrt{(1, 8^2 + 7, 8^2 \tan^2 \alpha)}$
Switzerland	1/25,000	$\pm(1 + 7 \tan \alpha)$
Switzerland	1/50,000	$\pm(1, 5 + 10 \tan \alpha)$
U.S.A.	1/50,000	$\pm(1, 8 + 15 \tan \alpha)$

Fig. 2.6. Table of formulae for calculation of errors in height of contours at various scales for some European map series. (After Imhof.)

Qualitative representation of height

The pictorial representation of the relief of the land has been attempted by cartographers from the very early days of mapping. In most primitive maps a general side view of a 'sugar loaf' type of hill was given, at first as separate symbols, sometimes arranged along lines and occasionally in an overall pattern resembling fish scales. An attempt at the depiction of the actual shape of the individual hills was made by Leonardo da Vinci on a map of Tuscany dated 1503 but one of the first efforts at producing a three dimensional impression of the landscape is that of Gygers map of the Swiss Canton of Zürich made in 1667 at a scale of 1/32,000.

The drawing of lines *down* the slopes of hills so that the lines are closer together or thicker at the top and more widely spaced or thinner lower down had been practised for a long time but the formal introduction of *hachures* following specific rules dates from the nineteenth century.

The following rules should be observed in drawing hachures:

(a) They must be drawn in the direction of maximum slope.
(b) They must be arranged in rows, not drawn down the entire slope.
(c) The length of each stroke should indicate an equal vertical drop in height. Thus a series of short strokes will be drawn over a steep slope, longer strokes will indicate a gentler gradient.
(d) The thickness of each stroke must be constant unless a form of 'illuminated relief' is being presented.

A modification of hachuring, known as the Lehmann system (Lyons 1914) keeps the hachure lines at an equal spacing but the angle of slope is shown by the thickness of the lines. A definite relationship exists between the two which on the Lehmann system gives a completely dark cover on slopes of 45° or more. Other systems allow for a

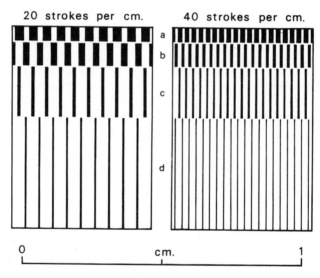

Fig. 2.7. Hachures, the Lehmann system in which width of line corresponds to slope angle for two different numbers of lines to the inch.
(a) Slopes of 30°. (b) Slopes of 20°. (c) Slopes of 10°. (d) Slopes of 5°.
Formula for width of hachure is given by the relationship

$$\frac{Width}{i} = \frac{x°}{45° - x°}$$

where x is the angle of slope and i is the interval between the lines. Thus slopes of 45° will be represented by solid colour.

black/white gradation until slopes of 60° are reached. Yet again a system of, say 4 'tones' allows for some flexibility. The number of strokes per centimetre also determines the apparent density of hachuring (fig. 2.7).

Hachures may be applied indiscriminately to all slopes irrespective of their aspect but a much better impression of the landscape can be given if shading is introduced. Here the precise relationship to slope angle is lost but if, for instance, slopes facing an apparent source of illumination in the north-west are shown with a lighter network of hachures than those to the south-east, an illusion of three dimensional relief is produced. This form of hachuring can be termed 'shaded hachuring' (fig. 2.8).

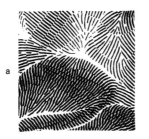
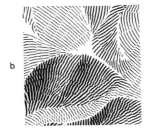

Fig. 2.8. Shading of hachures can be to represent angle of slope (a), or to indicate areas in shadow from a supposed light direction (b).

The method of using apparent illumination from the top left (north-west) corner is used in other methods of relief depiction. Apparent lighting from vertically above, so that valley bottoms are darkened and ridges left light in colour, has also been used but lighting from other presumed directions gives the impression of 'inverted relief'. An actual three-dimensional model of the landscape with an illumination from the top left will possess areas of light and shade which can be expressed in gradations of tone ranging from white to the full colour. If a colour plate is made either from a photograph of such a model or by the application of tones assuming the area to be lighted from one specific direction it is seen that the resulting tones grade smoothly to each other. Such tones occur in nature and are recorded in photographs but require presentation in printing processes by the use of 'half-tone' methods (see pp. 133–4). The resulting shading on the map is referred to as illuminated relief, plastic shading or hill shading and can obviously be shown in any desired colour. Examples of its use are in maps at scales varying from topographical sheets at 1/25,000 and 1/50,000 (e.g. French, German, Swiss series), 1/63,360 (Tourist sheets of the British one inch to the mile map) to small scale atlas maps produced in many countries.

Colour has been used in the depiction of relief for a number of years in the method known as 'layer tinting'. Such colour or shading is referred to as a *chorisogram* and is applied to the spaces on the map between the isograms which in this case are the contour lines. Common usage in atlases and wall maps is for a system of colours grading from light to dark as the land increases in height. It has become established practice to have a range of colours through the tints of green, brown and red, with purples and/or whites on the highest areas.

Few maps use only one system for the representation of relief as described above, combinations of both quantitative and qualitative methods being generally adopted. Even on the largest scale plans where precise values are given in the form of spot heights some additional symbols will be used for abrupt breaks of slope such as embankments or quarry faces. Here the hachure symbol or a modification or hachuring as in the cliff symbol, with sketches of rock outcrops, are used (fig. 2.9). Where a hill shading technique is used it is usually necessary to print this shading in a separate colour, but it is possible to combine this with layer shading.

A particularly interesting combination of layer tints and hill shading has been used in a map of the Freetown area at 1/50,000 produced by the Directorate of Overseas Survey for the Director of Surveys and Lands, Sierra Leone (Carmichael 1969).

Fig. 2.9. Symbols to represent steep irregular slopes where contouring is not possible. Modification of the hachuring principle of lines down the steepest slope (b), or of horizontal hachures or form lines (a) can be used separately, or combined as in (c).

Drainage lines

Lines of drainage are conventionally represented in the coloured topographic map by blue lines, bodies of water by a blue shade and information relating to water by lettering in blue. Other distinctions that are made relate to perennial and intermittent drainage (the latter usually shown with a broken line) and the change between riverine water and tidal water, i.e. the highest point to which tides flow often needs to be distinguished. Sources of water, springs, wells, etc. may or may not be shown, often depending on the map scale.

Blue is also the colour normally used by the cartographer for depicting details relating to ice, e.g. limits of glaciers, contour lines across glaciers, icefalls and crevasse features of the more permanent type.

This convention is, however, not always followed when the features of the seas and oceans are to be shown. As was stated earlier the datum line for all height information is usually taken as mean sea level. More significant lines on coastal margins are those representing high and low tide marks and, if the map scale permits, these can both be indicated.

Information relating to the shapes of the floors of lakes and oceans can be presented in much the same way as relief information but in most cases only submarine contour lines are used, layer shading being found in some atlases and 'hill' shading in relatively few specialised maps. The navigator's or yachtsmen's needs of maps of sea areas are somewhat different from the land-user and the mapping agencies serving them produce a specialised form of map – the hydrographic chart.

Limited hydrographic information is portrayed on most maps produced by topographic mapping agencies but is given in greater detail on specialist charts prepared by national hydrographic bureaux. Such charts fall into three main categories – ocean charts, harbour plans and navigational charts. The information shown on charts and plans will include details of water depths (in fathoms, feet or metres), nature of the sea bottom, position of reefs and hazards to navigation as well as data relating to tides and sea currents. Navigational charts are prepared on various map projections and generally have a close graticule network to assist with the plotting of position and courses. They can also be of an 'outline' variety covering a large area or may be designed for use with some navigational aid (e.g. Decca systems).

Hydrographic charts show depth of water by printing soundings as spot values and also by inserting fathom lines – isobaths – or depth contours. These are related to a datum stated on the Tidal Information Table on the chart, the actual soundings normally having been 'reduced' to the level of lowest low water. Where a sounding is given its precise position is that of the middle point of the figure(s). A different symbol is used for each fathom line and different print styles for soundings of different degrees of accuracy. A full explanation of the 'Symbols and abbreviations used on Admiralty charts' is published by the Hydrographic Department as chart no. 5011.

Non-coastal information will be given only where relevant for navigational purposes, e.g. prominent landmarks. Land relief is shown by hachuring, broken contours or form lines. In order that all navigational information is up to date a system of continuous revision is available through the daily publication of 'notices to mariners'. Chart-users are required to make any necessary amendments to their charts from the information in these 'notices'. Where considerable changes occur the chart is withdrawn and a new issue produced. Publication runs of hydrographic charts are therefore small and stocks kept at low levels.

Until recently, charts were produced in a nearly monochrome style, the only colour being small patches of red or purple used for lighthouses and similar navigational aids. International specifications now include colouring, e.g. yellow for land areas, and layer colouring to indicate the shallower water over continental shelf regions. This may be in one or more shades of blue and blue is also used to indicate marine contours (isobaths). The metre is now the international unit for depths and the fathom unit is disappearing from use (fig. 2.11).

Most British charts were formerly prepared from engraved copper plates and in recent years have been photographically transferred to zinc-enamel for offset lithographic printing. The 'engraved' style is therefore still present in such charts and is particularly noticeable in the profiles added to give an indication of the mariner's view of the coastline (fig. 2.10).

Tresco Island

Star Castle
Hangman I.

Bryher Island   Shipman Head

Fig. 2.10. Extract from Hydrographic Chart of the Isles of Scilly to show additional illustrative sketch to assist in the recognition of harbour entrance courses.

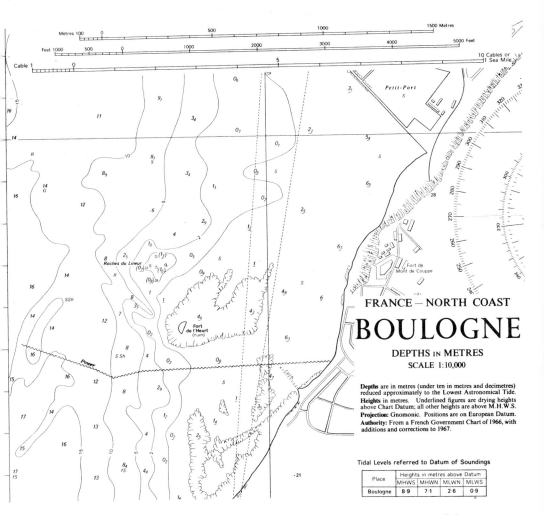

Fig. 2.11. Extract from Admiralty Hydrographic chart of Boulogne. Colour washes distinguished water depths below 5 metres, tidal flats and land areas. These appear as grey tones on this reduced scale (about 1/17,000) photocopy.

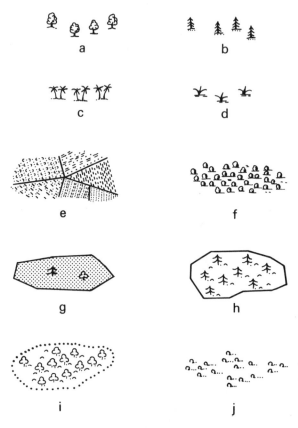

Fig. 2.12. *Some vegetation symbols used on maps and charts.*
(*a*) *Deciduous trees.* (*b*) *Conifers.* (*c*) *Palm trees.* (*d*) *Nipa palms.* (*e*) *Culti-vated fields.* (*f*) *Bush growth.* (*g*) *Parkland with some tree growth.* (*h*) *Fenced coniferous woodland.* (*i*) *Unfenced deciduous woodland.* (*j*) *Scrub undergrowth.* (*a–f*) *from Admiralty charts*, (*g–j*) *from Ordnance maps.*

Vegetation

The mapping of vegetation distributions necessitates areal symbols of some form (McGrath 1965, 1966). The requirement can be met by means of a repeated single symbol, e.g. that representing a tree or bush which on the very largest scale of map will be a precise locational representation of individual trees, etc. As the map scale reduces, the exact positioning of each object will not be possible because of the size of symbol used, since the symbol must remain large enough to be easily recognised (fig. 2.12). A generalised pattern of symbols will then be used or, alternatively, shading or colours to represent the distribution of the vegetation under study. These are limited on most topo-graphic map series to a few broad categories of natural or semi-natural vegetation such as heathland, rough pasture, woodland. Shading, symbols etc. may also be used to represent distinctive cultivated land use categories such as plantations or orchards. One feature of the British landscape which has given rise to a symbol which may be considered to belong to the 'vegetation' category is that used for park-lands, both public and private – which are shown by stipple on Ordnance Survey maps of 1/250,000 and larger scales (fig. 2.12).

3 Representation of the activities of man

The mapping of all topographic details of the earth's surface leads to problems of representation and of generalisation. These are a result of the different types of man's *occupance*, e.g. agricultural land use, settlement, communications, and also stem from different *scales* of map representation. Cartographic problems arise, too, in representing information on both single-purpose thematic maps and on multi-purpose topographic maps. One must consider such matters as statistical representation, colour choice, balance of tones, use of symbols, lettering, etc.

In all cases, however, the information depicted falls into one of the following categories:

(a) LINEAR information, e.g. boundaries or communication lines, representing either dividing lines or flows and movement.

(b) AREAL information, which usually leads to map shading or colour tones, e.g. individual buildings at large scales, entire settlements at small scales, land use information at all scales. This is also referred to as 'patch' data.

(c) SYMBOLISED information, which can be information relating to a POINT where a conventional shape represents a feature in a pictorial or other manner, e.g. geometrical shape, or in the form of a statistical diagram, bar-graph, divided circle, etc. relating either to a point or to an area.

(d) WRITTEN information in words, abbreviations, numbers, etc. This can relate to a point, area or line.

All four categories indicate a representation of the topography in some 'conventional' form and lead to the necessity of a map *legend*. This is sometimes referred to as the map 'key' but this term has other technical connotations in cartography, e.g. the base outline of the map before colour separations are prepared in map reproduction (chapter 12).

Map information can be presented in any of the above methods either generally, merely to give qualitative information as to location or activity (e.g. town name, type of industry), or more precisely. The latter will give quantitative information and factual data (e.g. town population or industrial production for a given date).

Another factor common to all the categories is that the information shown on the map may not be physically evident in the topography. Thus a pipeline may not yield a continuous line on the ground, and for this reason may not have been mapped by the land surveyor. Place name information, administrative boundaries, are further examples. Lettering or explanatory words may well be required to supplement or to explain other symbols on the map (e.g. evidence of former land use or occupance – prehistoric settlements and burial sites, forts and tumuli).

The cartographer will aim to depict as much of the ground information as possible at the scale to which he is working and will need to balance possibly conflicting needs of the map's purpose, map symbols, written information, etc. Over many years a number of particular symbols and colours have become accepted or 'conventional' usage but there have also been changing fashions in map representation stemming from advances both in the complexity of the topography and in cartographic techniques or processes of reproduction.

Linear information

Printed lines on the surface of the map represent a whole series of topographic features. However, lines can generally be taken to represent boundaries in some form. Where such boundaries are clear and definite they will be shown by a full line, e.g. fences, walls, edges of buildings. Where low and indefinite, e.g. road/pavement edges, unfenced fields, they will be given a broken line. Thus on the largest scale maps a road will be indicated by full lines for the hedges, together with broken lines for pavements. Two printed lines will be present also for roads at smaller scales, probably with a colour fill between them but one or more of the lines may be broken to indicate lack of fencing, or road category. Eventually, however, a single line of colour may be used to represent the road or other communications line. Even without the use of colour it is possible to utilise a wide range of line symbols by introducing broken lines, chain lines or additional symbols along or across the line (fig. 3.1).

All these examples, whether for roads, railways, canals, pipelines, transmission lines, administrative boundaries or even for distinction between hedges, wire fences, boundaries with trees, are of a qualitative or purely descriptive nature.

In order to introduce a quantitative element in a line symbol it will be necessary to provide it with a dimension. At its simplest this can be a written figure alongside the line but on thematic maps showing flows along route lines a variation in line-thickness is the usual practice (fig. 3.2). In cases where a large variation in values is to be catered for, it may be necessary to use a pictorial device whereby the line becomes an apparent three-dimensional shape such as a cylindrical pipe or one of rectangular cross-section.

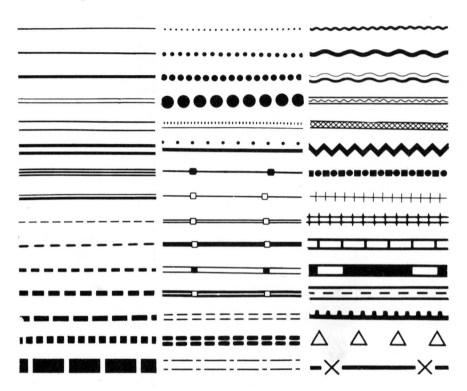

Fig. 3.1. Line symbols. Examples of different symbols which may be used for lines, e.g. boundaries, communication lines, flow-lines. Most of these are available in pre-printed form, as tapes which can be applied to drawing surfaces (p. 123): the possible range of linear symbols is far more extensive than can be shown here.

Fig. 3.2. Flow line maps. The two map examples given here show respectively (top and bottom) domestic weekly passenger air traffic (1966) and main line rail freight (1965) in South Africa. The width of the line is proportional to the value and thus requires a legend in addition to the circle scale for airport passenger handling in the top diagram. Less important connections are not dimensioned. Note that in the lower diagram the flowlines are following the actual routes and vary along their course. (After Board et al., 'Regional Studies' 1970.)

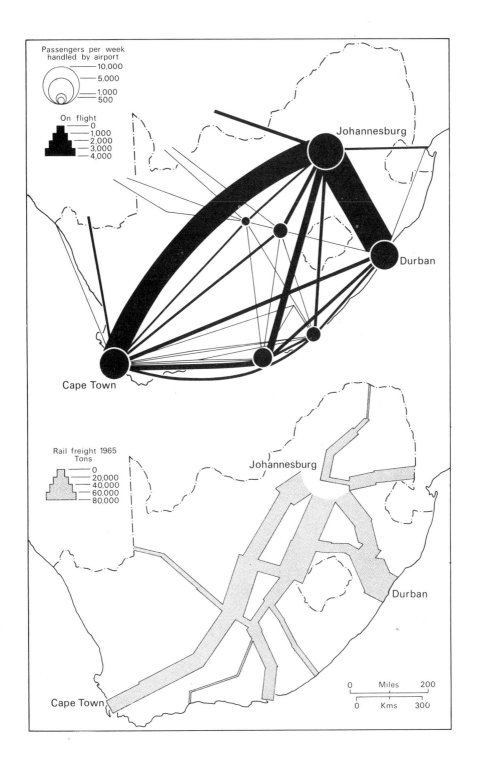

Passengers per week
handled by airport
10,000
5,000
1,000
500

On flight
0
1,000
2,000
3,000
4,000

Johannesburg

Durban

Cape Town

Rail freight 1965
Tons
0
20,000
40,000
60,000
80,000

Johannesburg

Durban

Cape Town

Miles
0   200

Kms
0   300

Areal information

Many activities in the field of human geography can be considered under this heading, i.e. settlement, industry, agriculture, actual land use, potential development. The larger the scale of map at which the topography is to be represented the more easily can such information be presented by the scale plan form of the activity, e.g. individual houses, industrial sites and land parcels. Particular uses or activities can be shown by a shading or colour placed over the area concerned. Some examples of the various shadings used on maps are given in fig. 3.3. This shading can indicate the activity in a purely descriptive way as on land-use maps or it can give numerical information according to a pre-determined shading or colour legend. Such shading/colours are referred to as *chorograms* and where the shading/colour is applied to a statistical or administrative division the term *choropleth* is used.

The term *chorisogram* has already been defined (p. 24) where shading is applied to the intervals between measured isograms (lines along which a value is constant). In thematic mapping other lines may be drawn – merely to join points having the same value and these are called *isopleths*. Shading between such lines is therefore technically referred to as *chorisopleth*. Similarly if the iso-line are accurately measured they should be termed isometric lines and the shading between them would be referred to as a *chorisometer*.

Problems which occur in any such quantitative representation of data can be considered under the headings of the data itself and the choice of shading. Problems provided by the data generally stem from the sources from which it comes. In the published thematic map it can be assumed that the data is accurate and adequate and the only problem is that of interpretation. The greatest importance therefore lies in the map legend. Fig. 3.4 shows two methods of representing the same data; in the first case no account is taken of the statistical distribution of the values whilst in the second the shading refers to one or more standard deviations from the mean. Alternatively, these figures could have been mapped by reference to the position of the mean, upper and lower quartiles, etc. A discussion of these particular examples is to be found in the *Cartographic Journal* (Storrie and Jackson 1967) and there is a further discussion of population mapping together with a number of maps in *Transactions of the Institute of British Geographers* no. 43, 1968 (Hunt 1968).

Symbolic representation

The simplest methods of representing distributional data are those of dots and proportional geometrical shapes. The dot is clearly the simplest symbol that can be used and, basically, one dot can represent *one occurrence* (single or group) of the data to be mapped. Usually the single

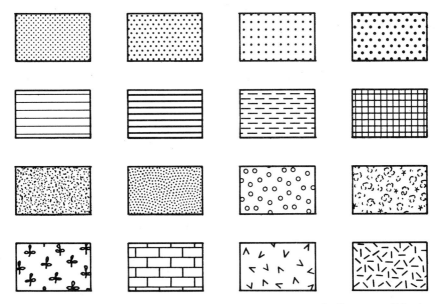

Fig. 3.3. A small selection of the many monochrome shadings possible for cartographic use. All the examples shown are from pre-printed sheets and the top two rows show possible ranges which could be used against a numerical scale. The proportions of dark to light can be related precisely to given values if required, by varying line widths or dot sizes and spacings. The bottom two rows show a variety of area shading symbols which could be used as choropleths.

dot will stand for a number of items and the dot itself may be replaced by some other simple shape either geometrical or pictorial. Ideally it should be possible to count the number of symbols and so obtain the value mapped. The distribution of these symbols can be made in a uniform manner – as for vegetation information, described on p. 28. It is more usual, however, for the symbols to be placed in geographically correct positions, e.g. dots for population distribution will be clustered in urban localities and scattered over remote hilly areas. Such plotting of a distribution is referred to as *dasymetric plotting.* Once the data to be represented exceeds that which can be satisfactorily shown by a number of single symbols it is customary to use a geometrical figure which can be measured against a given scale. The most usual figure is a circle and in order that this may correctly represent the required data the *area* of the circle will be made proportional to the figure it is to show. The legend to such a thematic map will then include a scale as in fig. 3.5 – against which the map-user can measure the symbol by means of a pair of dividers.

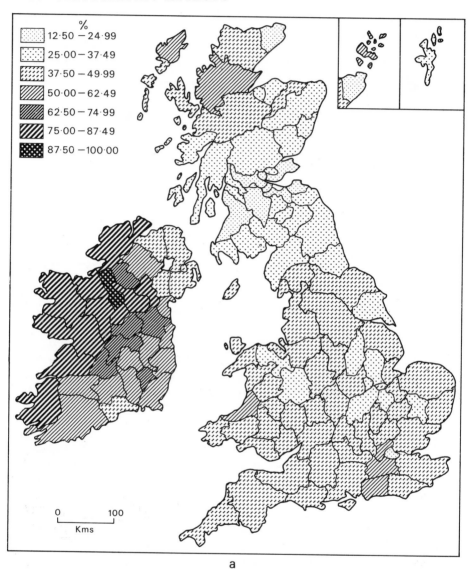

%
12·50 — 24·99
25·00 — 37·49
37·50 — 49·99
50·00 — 62·49
62·50 — 74·99
75·00 — 87·49
87·50 — 100·00

0   100
Kms

a

Fig. 3.4. Representation of statistical data. These two maps present the same information, the proportions of owner-occupiers by counties in the British Isles, in two different ways. In (a) the data is that of actual percentages, with equal divisions of the 0–100 scale (no values below 12·50). Note that there is no overlap of value, i.e. the divisions are 12·50–24·99, 25·00–37·49 etc.

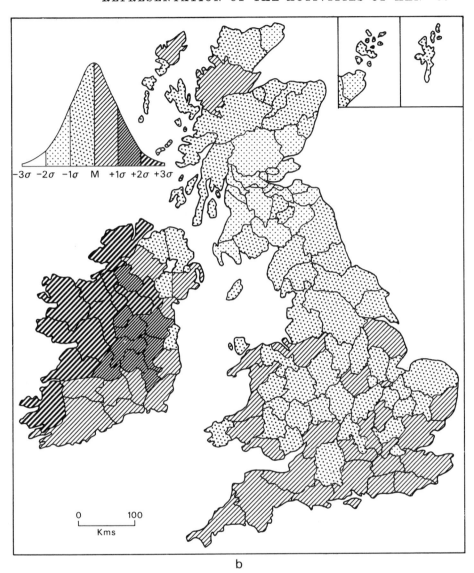

b

(b) presents the same data in terms of deviations from the mean, the range here being across five standard deviations. Shading has been applied to both these maps such that there is a distinction between values below and above the mean value (44·52%), and in (a) the category containing this figure is shaded distinctively.

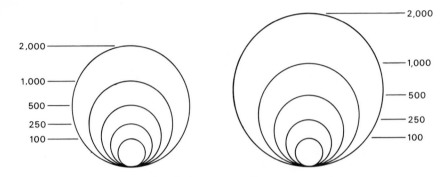

Fig. 3.5. Proportional circles. When circles are drawn to represent quantitative data it is usually regarded as satisfactory to make the area *of the circle proportional to the data (left-hand diagram). However, the impression of relative sizes is not entirely satisfactory and it is better if an enlargement is made (right-hand diagram). A table providing values for this calculation is given in Robinson and Sale (1969, Appendix F).*

Written information

By far the easiest way in which to record and present factual information on the map is by means of a printed statement in words and/or figures. The simplest and most obvious examples are place names and values for ground height. Provided the map scale is large enough, very much more information can be recorded in this way. Indeed, the field surveyor of, for example, agricultural land use, will probably proceed by annotating the large scale plan. For complex patterns, such as those in urban areas, it may be necessary to record only reference numbers on the map and list details of building, occupance and other data on a separate booking form.

Most names on published topographic maps are place names. In order to distinguish between places of different size or of different status (i.e. administrative function, capital cities, towns, parishes, etc.) a different size and style of lettering can be employed (fig. 3.6). Also, variations in lettering can be used to indicate different historical ages of the topographic features. In order to overcome the extravagant use of space on the map surface which lettering can involve, abbreviations are frequently introduced. These will usually refer to a given location, normally marked with a point and the full legend to the map will include a comprehensive list of all these abbreviations (fig. 4.1 and Appendix II).

Numbering on maps includes such information as spot heights and, on large scale plans, reference numbers to the individual land 'parcels'.

(a) POPULATED PLACES ⎯⎯⎯⎯⎯⎯⎯ ○ ┼┼━━

Over 500,000⎯⎯⎯⎯⎯ **LOS ANGELES**
100,000 to 500,000⎯⎯⎯⎯⎯⎯⎯⎯**OMAHA**
25,000 to 100,000 ⎯⎯⎯⎯⎯⎯**GALVESTON**
5,000 to 25,000 ⎯⎯⎯⎯⎯⎯⎯⎯**Laramie**
1,000 to 5,000 ⎯⎯⎯⎯⎯⎯⎯**Grand Coulee**
Less than 1,000 ⎯⎯⎯⎯⎯⎯⎯⎯ Sun Valley

Fig. 3.6. Examples of lettering styles and sizes used to indicate different features on maps.
(a) Quantitative use, where size of lettering indicates approximate size of detail, e.g. town size by population – extract from U.S. 1/250,000 map legend.

(b) Qualitative use, where different printing styles indicate different features – extract from Conventional Signs sheet for Ordnance Survey 1/2,500 County sheet lines.

Parish Churches & Villages

Other Villages

PARKS & DEMESNES

Gentlemen's Seats

Manufactories, Mines, Farms, Locks

Local Authority Establishments

Bridges *(On Main Roads),Bridges (Other)*

Isolated Houses

BAYS & HARBOURS

NAVIGABLE RIVERS & CANALS

Small Rivers & Brooks

BOGS, MOORS & FORESTS

(b)   Woods & Copses

On British Ordnance Survey maps at 1/2,500 and 1/1,250 scales these parcel numbers are quoted together with a statement of the area of the parcel (fig. 3.7). Various methods of presenting this information have been used and these are summarised in the O.S. publication *Parcel Numbers and Areas on 1/2500 Scale Plans.*

Other numbers which can occur on topographic maps are those relating to population (viz. French 1/25,000 and 1/50,000 series figures given after the printed place name) and such information as distances along roads between marked points, and the official road classification number. In the map margins distances to settlements on adjoining sheets are often given.

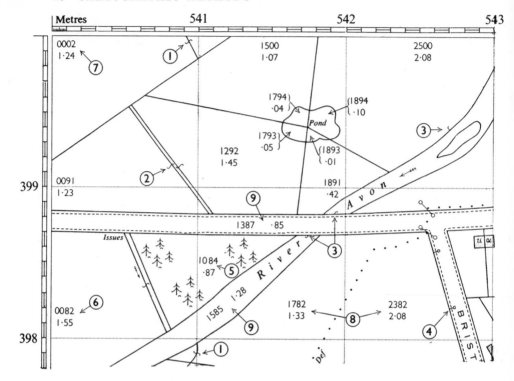

Fig. 3.7. Parcel numbers and areas on Ordnance Survey 1/2,500 plans.

1. Bracing symbol to indicate areas combined.

2. Centre bracing symbol to indicate that area is measured to the centre of this strip.

3. Open bracing where parcel concerned is crossed by a second parcel (road in this case).

4. Symbol to indicate the edge of the built-up area. This latter contains a large number of small parcels, difficult to show with individual areas, therefore all summed within this built-up portion.

5. Reference number (grid reference to ten metres of approximate centre of parcel) and area – in hectares on metric maps, acres otherwise – of parcel defined by surrounding boundary lines.

6. Parcel number and area to edge of map.

7. Corner parcel number and area where parcel extends beyond map corner.

8. Parcels in separate parishes.

9. Parcels are sections of road and river.

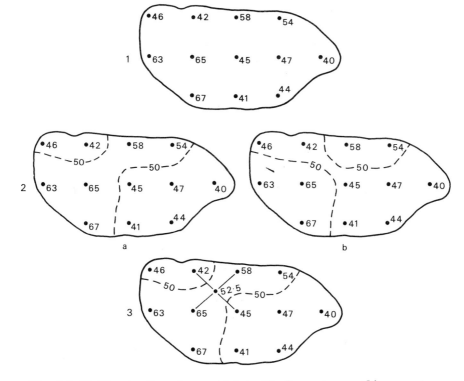

Fig. 3.8. Plotting iso-lines from grid data (1). In most cases this presents no problems but in this particular example two solutions (2) seem possible. If the four surrounding values are averaged (3) a value is obtained for the central point and this provides a solution or alternatives. The position of the iso-lines in the remainder of the example is obtained by assuming a steady change in value between any two points. (Board 1967, after Mackay 1953.) There are cases, however, where it is necessary to choose between the two possible solutions on grounds of the information to be conveyed. Thus, if the appropriate map were to be used for air navigation, where ground heights are critical, a solution should be given indicating the highest possible value from the data (i.e. solution 2a is to be preferred). On the other hand if the values relate to critical depths, as on a hydrographic chart, a solution which shows the minimum depth is preferable, i.e. 2b.

Distributional information

Thematic maps can present geographical and distributional information in the variety of ways considered above but it is often necessary on such maps to show or examine statistical data for an area rather than for isolated points. In addition to the choropleth methods or the presentation of figures and symbols on the map the cartographer may use the method allied to contouring for the depiction of relief. This is the technique of drawing *iso-lines*, lines joining points of equal value. In contouring, the actual land surface can be measured and the precise position of the contour so obtained. In other distributions it is normally only possible to know precise values at a network of points. These points may be randomly positioned (random in either a literal or statistical sense) or regularly placed according to some network or grid. In either case it is likely to be necessary to interpolate the position of the isoline from a network of spot values. An assumption must be made as to the rate of change of the data between the given values and in most cases this will be taken as being at a uniform rate. A method of calculating the position of iso-lines for given spot values is shown in fig. 3.8. The drawing of iso-lines and also the plotting of choropleths can be undertaken by various computer programs. Many other ways of representing distributional data are now available as a result of computer mapping systems (Rosing and Wood 1971) or from data banks (Hägerstrand 1967).

Part II

Map types and map series

4 Map types

The distinction between two major map types – topographic and thematic – has already been made. Published map series, however, fall into a number of distinctive types which reflect the basic mapping need and also the 'house style' of the map-producing agency. This style is a result of the particular specifications to which the map publisher has worked and is due to the combination of controls of layout, lettering, colouring and additional information, as well as the topographic and/or thematic information portrayed on the map itself. The method of reproduction provides a further factor in establishing the map style.

Distinctive map styles characterise not only the two main groups of maps so far defined but more particularly the organisation producing the map. The official map series of different countries thus show variations ranging from the shape and size of the map sheet to the way in which some topographic features may be presented. Careful reference to the map 'legend' must therefore be made. The legend printed on the map itself may not be the full legend adequate for the entire map series (e.g. British, Swiss 1/25,000 maps) and in such cases it will be

Fig. 4.1. Extract from Dutch map legend. A small portion of the legend to the Dutch official topographic map series at 1/50,000 scale to show how a multi-lingual legend can be set out.

Lighthouse
Phare | *coordonnées connues*

Merksteen *Rijksdriehoeksmeting (RD)*
State Survey stone *(RD)*
Borne de la *Triangulation Nationale (RD)*

a Gemeentehuis, Municipal hall, Mairie
b P.T.T. kantoor, P.T.T. office,   Bureau de P.T.T.

a Kapel,   Chapel,   Chapelle
b Kruis,   Cross,   Croix
c Wegwijzer, Sign-post, Poteau indicateur

a Windmolen, Windmill, Moulin à vent
b Watermolen, Watermill, Moulin à eau

a Windmolentje, Small windmill, Petit moulin à vent
b Windmotor,   Windmotor,   Aéromoteur

Gemaal
  a stoom
  b motor
  c elektrisch

Pumping-engine
  a steam
  b motor
  c electric

Épuise
  a à vapeur
  b à moteur
  c électrique

a Oliepompinst. Oilpumping unit, Pompe de pétrole
b Seinmast,   Signalpost,   Poteau de signal
c Gedenkteken, Memorial,   Monument
d Hunebed,   Cairn,   Dolmen

necessary to obtain the separately published legend. In many European cases the legend is multi-lingual, e.g. French, Dutch and English of the Belgian and Dutch 1/25,000 series (printed on the map), French, German and Italian of the Swiss map series and the Russian and English of some Russian maps, particularly the more readily available small scale maps in this last case.

Official map series

The official topographic map series of any country, although with differing styles, usually form a distinctive group in which both monochrome ('outline') or fully coloured editions are available depicting relief, settlement, communication and some vegetational information as well as names and political boundaries. The larger scale maps prepared for cadastral purposes are usually in a single colour, black or grey line work being the most common. Planning maps, on the other hand, are more usually produced with colour shading. Both these categories are the work of official mapping bodies and bear the distinctive style of that body. As scales increase the monochrome cadastral maps become more and more alike in appearance but the thematic maps of planning bodies often reflect the stamp of the printers and publishers.

Commercial maps

Maps produced by non-official agencies usually rely heavily on the various official topographical map series. The maps which such bodies subsequently produce are often of the thematic type and fall into such categories as:

1. Town plans at various scales.
2. Road maps at various scales.
3. Itineraries – or maps of particular routes – and location or site maps.
4. Tourist maps of specific regions.
5. Miscellaneous thematic maps, e.g. marketing, distributions, productions, etc.
6. Network maps of rail, air or other communication patterns.

Nevertheless, commercial bodies also produce topographic maps, sometimes as individual sheets and sometimes as a series giving national cover (e.g. Bartholomew's 'Half-inch' maps).

Sea and air charts

Maps which fall somewhere between the specialised thematic maps listed above and the official topographic map series are those relating to sea and air communications. Most maritime countries produce their own hydrographic charts (p. 25) not only of adjacent coastal areas but also covering overseas waters so that sailors may have a chart printed in their native language. Many of the latter are derived from the work of other hydrographic surveys.

For aeronautical purposes a further range of special maps or charts are required, and here the special problem is that of 'up-to-dateness' as restricted areas, flight lanes, etc. may be subject to changes at short notice. Overprinting of symbols on a standard topographical map is a possible solution to this but the normal method is the production of a special 'air' edition making use of standardised symbols, shadings, etc. as laid down by the International Civil Aeronautical Organisation (ICAO).

Both hydrographic and aeronautical map-users also require specialised charts relating to radio and other navigational aids. Here the usual practice is to utilise a base map with essential outlines on to which is printed the appropriate grid lines of the radio system, e.g. Decca, Loran and Consol charts.

In the succeeding chapters the official map series and some of the more widely available commercial maps produced in both the United Kingdom and some Western European countries will be briefly considered.

Aerial photographs as maps

Aerial photography has been used for several decades in the preparation of topographic and other maps but usually as a survey method through the principles of photogrammetry. The air photographs themselves can, however, be substituted for the map in a number of ways with varying degrees of precision. For rapid mapping purposes a set of photographs may be approximately placed in positions controlled by a skeletal survey framework. This arrangement of photographs can be itself photocopied and produced as a *'print laydown'*.

Alternatively the photographs can be more carefully joined and again, controlled by a skeletal survey framework, copied in a *mosaic* form. These mosaics may exist in both an unrectified and a rectified form. The latter indicates that correction has been made not only for the various distortions due to aircraft tilt, etc. but also for scale.

No allowance can be made, however, for differences due to height variations on the photographs and, unless the area considered is perfectly level, the scale will vary with height. This can be overcome by the methods of *orthophotography* producing an *orthophotomap* (Scott 1969 and Zuylen 1969).

The photograph, consisting of various tones, can also be broken down into various component parts of light and shade of which the darkest tones are the shadows which outline areas. A series of masks and screens can be used to differentiate these parts of a photograph and the

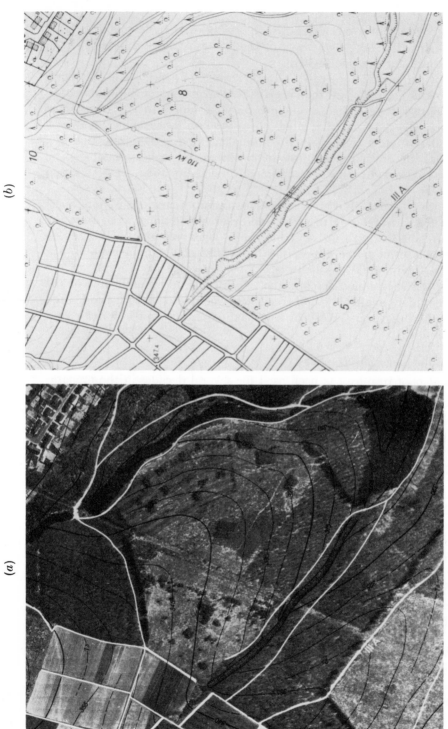

Fig. 4.2. Example of orthophotomap. This section of the German orthophotomap (a) is at a scale of 1/5,000 and shows contours in black over-printed on the scale-corrected and rectified air photograph. Roads and tracks outside built-up areas have been made blank, spot-heights, road numbers and contour labelling have been added. (b) is an extract from the conventional map of the same area.

resulting process is known as *Pictomap* (Photographic Image Conversion by Tonal Masking Procedures – U.S. Army Map Service).

These processes lead to the production of maps in which the air photograph base is clear to a greater or lesser extent and which can be termed *photomaps*. These are produced at a variety of scales, examples ranging from 1/5,000 in Nordrhein-Westfalen to 1/50,000 in Saudi Arabia. In most cases some interpretation is made by adding linework to the base photographs – to show contours, emphasise roads, houses, boundaries, etc. Names, too, must be added and separate tones or land use interpretations can be put on by the use of colour – as with the Swedish Economic Map at a scale of 1/10,000.

Computer maps

Maps prepared by computer are becoming increasingly available and there are a number of computer programs which may be used to present geographical and other data in diagram form. Some of these programs have been prepared for particular purposes: others can be used with a variety of data or for various methods of analysing data and presenting the results in map form.

Maps produced by computers are generally prepared directly on the lineprinter output sheet. This is a continuous roll of paper which is folded into pages approximately 390 mm $\times$ 280 mm. The computer prints one line or row of characters at a time – hence the term lineprinter – and the characters are those of the conventional typewriter keyboard. Overprinting of characters allows for the printing of symbols of different intensity (Robertson 1967 and Rosing 1969). Each character occupies an area $\frac{1}{10}$ in in the X axis and either $\frac{1}{8}$ in or $\frac{1}{6}$ in in the Y axis (fig. 4.3). Each symbol is printed within this rectangle and a narrow margin of blank paper will result between each symbol giving a network of white lines over the map. Note that this shape is *not* rectangular as is the normal map grid. Linear maps can, however, be produced from computer data in co-ordinate form (fig. 10.3) (Diello 1969).

Amongst the programs available for the preparation of maps by computer are:

1. SYMAP (Synagraphic Mapping System; Harvard University).
2. LINMAP (Ministry of Housing and Local Government; London).
3. COLMAP – a colour map system based on LINMAP.
4. MAPIT – a program for the production of flow maps, dot maps and graduated symbols.

SYMAP can be used with data relating to given co-ordinates to prepare either an isopleth or choropleth map as required.

LINMAP was developed in order to operate a 'data bank' of given statistics relating to the General Register Office Census information at

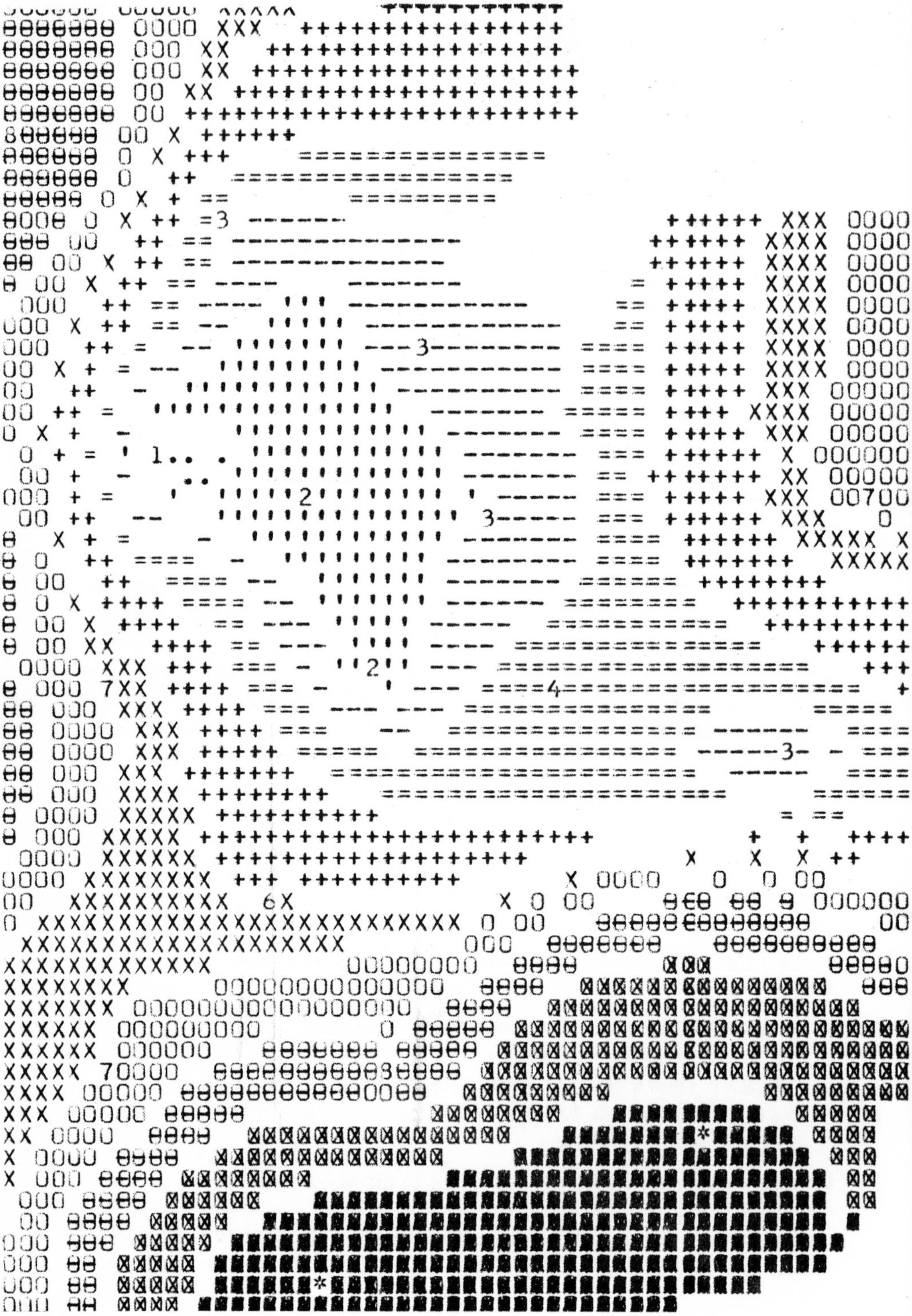

ward and parish levels. The maps so produced can be either dotmaps, grid maps (shading related to a specified grid pattern on the map), zone maps – a form of choropleth map and 'surfmaps' (from surface mapping) – in which values are calculated by interpolation from the sample values provided (Gaits 1969). The relationships between the National Grid system and the computer map are clearly exemplified by the Gridmap output which is in a 5 × 3 or 10 × 6 format and which must be made to conform to a square grid as shown in fig. 4.4.

The basic data for computer mapping is related to a rectilinear grid such as the Ordnance Survey national grid. This data is then stored on tape and can be used to prepare a key map on a grid *square* basis (unlike the normal computer output described above). In the case of COLMAP the key map is replaced by a series of separate colour plates for litho printing in a conventional manner (*Penrose Annual* 1971).

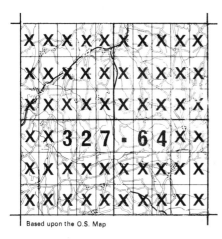

Based upon the O.S. Map

Fig. 4.4. Grid pattern used by Gridmap *computer program. The main squares conform to those of the National Grid, but these are subdivided into rectangles to conform to the computer output (after Gaits 1969).*

Fig. 4.3. Detail of print-out from Symap program showing arrangement of squares in lines (8 per inch vertically) and columns (10 per inch horizontally). Shading variations are produced by over-printing of type characteristics but because the ordinary type style is used, small white lines surround each rectangle.

5 British map series

There are many different map types produced in Great Britain for both general and special purposes as well as probably an even larger number of individual maps of specific areas. These may be conveniently grouped into two categories – in both of which will be found examples of topographic and thematic map types.

1. Maps produced by official mapping agencies, Government bodies, etc. such as the Ordnance Survey.
2. Commercial mapping firms. These produce maps of many overseas areas as well as of Britain.

In general, maps produced by the second group in Britain rely on base data from official surveys but for overseas work may undertake their own surveys.

The Ordnance Survey

The Ordnance Survey was established at the end of the eighteenth century by the detailing of military personnel under Major-General Roy to undertake a 'Trigonometrical Survey' of Great Britain and Ireland. For defence purposes this work was under the charge of the Board of Ordnance – hence the name of the survey. It was eventually transferred to the Office of Works and in 1890 to the Board of Agriculture, but since 1967 has been the responsibility of the Ministry of Housing and Local Government (since 1970 under the Department of the Environment).

The original offices of the Ordnance Survey were established in the Tower of London in 1791 and the first published topographic sheet – the one inch to the mile map of the county of Kent – appeared in 1801. Larger scale maps were produced later in the nineteenth century, starting with a series of six inch to the mile maps of Ireland and later of Great Britain. The first maps at the scale of twenty-five inches to the mile appeared in 1853. The early large scale mapping of Great Britain continued at these two scales, the larger one being used only for the major built-up areas. However, these maps were on a county by county basis both as regards their projection and sheet lines. Problems of discontinuity between counties or blocks of counties were considerable and the Davidson Committee on the Ordnance Survey reporting in 1938 recommended a number of changes. These were not fully implemented until the period after the Second World War and the final pattern of mapping is described below. Changes due to the impending introduction of metric units during the 1970's and 1980's are also noted.

Ordnance Survey map series

The largest scale topographic maps of Britain usually available are at the scale of approximately 50 inches to the mile, the representative fraction being 1/1,250. These maps are produced for towns and urban areas with a population of about 20,000 or more and are square in format, each sheet covering an area of 500 sq in. The sheet lines are based on the squares of the National Grid.

The old county twenty-five inches to the mile maps have been superseded by maps at a scale of 1/2,500. These will eventually cover the whole of Great Britain apart from areas of mountain and moorland. Some of these map sheets have been produced in a rectangular format covering an area on the ground of 2 kms from west to east by 1 km north–south and this will be the eventual pattern for all maps in this series. The information shown on both these maps is purely planimetric together with spot heights and heights of bench marks. In metric units the former are to one place of decimals and the latter to two decimal places. Areas are quoted at present to two decimal places of an acre but with metrication this will be changed to three decimal places of a hectare.

Maps at a scale of 1/10,560 (six inches to the mile) have been published for the whole of Great Britain. National Grid sheet lines are used for the whole of England and Wales as well as most of Scotland but for remote highland areas it may still be necessary to consult the original county sheets which may not be contoured. Nevertheless, maps at the six inches to the mile scale are the largest for which a considerable amount of relief information can be obtained, the contour interval varying from 25 ft on the fully revised sheets to 50 ft on sheets of a 'Provisional Edition' which was compiled by re-drawing from the original county maps together with additional revisions. After metrication this series will appear at the scale of 1/10,000 with contours at 10 m intervals in the more mountainous areas, 5 m elsewhere. It is likely to be many years before the entire country is covered at this new scale and a number of different map series will therefore co-exist for some time.[1] All the above maps are based on the largest scale field surveys of the appropriate areas and have much in common. Generalisation is needed only to a very limited extent and is mainly confined to

1. The Ordnance Survey policy of 'continuous revision', however, will ensure that any major changes in the landscape will be recorded and reasonably up-to-date maps will remain available.

roads and building outlines at the 1/10,560 scale. The result of such 'widening' of roads, however, is to make inaccurate any measurements of area by geometrical methods. The contours on this map are produced photogrammetrically and detailed height information is confined to levelled spot heights, quoted to the nearest foot or one decimal of a metre. For greater information relating to heights at this scale it is necessary to turn to the printed lists of bench marks which are prepared for every 1 km square of the National Grid.

At smaller scales only one series at present is confined to exact large National Grid squares. This is the 1/25,000 series, popularly called the 'Two and a half inches to the mile' map, each sheet of which covers a ground area 10 kms by 10 kms. (The second full edition of this series, however, will be in a rectangular format of 20 kms west–east by 10 km north–south.) Colour printing makes its first full appearance at this scale as apart from brown or red contours on various six inch to the mile maps, monochrome black is the dominant colour of the larger scales. Four colour printing is used on the 1/25,000 maps of the first edition[2] which show most land parcels, contours at 25 ft intervals (to be 5 km and 10 km in lowland and upland areas respectively with metrication), generalised building shapes and woodland. Brown is used for the contours and road filling, blue for water, green for vegetation information and black for buildings, field boundaries, etc. A fine black hatching provides a tone over buildings and the grid lines are also in black.

The one inch map
As well as being the first map type produced by the Ordnance Survey this is probably the most well-known and no doubt forms the popular image of 'Ordnance maps'. Until recently the one inch to the mile map appeared in successive *editions*, each of which had distinctive production styles. Thus the First edition was a monochrome black in which relief was depicted by hachures. This was followed, in 1840, by maps of the 'New series', based on the six inch to the mile survey, and ultimately in 1893 by the Third edition. Different arrangements of the map sheets (or *sheet lines*) were used for these editions. The layout of the Third edition remains today (in its small sheet lines version) in the sheets of the one inch to the mile map of the Geological Survey (New Series) where it has been used in a black outline form as a base map.

The Fourth edition began in 1919 on different sheet lines and like the Third edition was a fully coloured map. The Fifth edition, which was started in 1928, marked a considerable change in style, particularly in lettering and also in the introduction of additional methods of relief

2. 'Provisional edition' maps at 1/25,000 are produced only in black, brown and blue – with grey in some earlier printings.

representation – hill shading in particular. This map series was never completed and when the need arose for maps during the Second World War the Fourth edition was revived in the form of 'War Revision' series. From 1946 onwards the sheet lines of the one inch to the mile map became more precisely related to maps at other scales by making these lines follow grid lines of the National Grid. This map was referred to as the Sixth edition and also as the New Popular edition with National Grid. It was confined to England and Wales, Scottish maps being separately covered by the Scottish Popular edition based on the older Fourth edition.

The Seventh *series* one inch to the mile map first appeared in 1959. This is basically on the same sheet lines as the Sixth edition but includes the whole of Great Britain. However, there have been a number of minor changes in style since its original publication. Some of these were concerned with the number of colours used, which have fluctuated from ten down to the present eight. Different editions within the series occur for each sheet, depending on the state of revision. For example, new roads and motorways may be added, or information relating to public rights of way over footpaths and bridleways may be made available. This will involve some modification to the map which will be recorded in two ways. Firstly, there will appear a reference to the change in the footnote outside the map margin such as: 'Reprinted with the addition of major new roads, 1970.' Secondly, an edition letter is also given to each revision of the map and this is shown in the bottom left-hand margin by 'A', 'B', etc. Such edition letters and footnotes are not confined to the one inch to the mile map; the appropriate edition letter is printed both in the top right-hand and the bottom left-hand margin of maps at larger scales.

With the advent of metric units a number of changes in the Ordnance Survey maps will probably include the disappearance of the one inch map.

Other government map-producing agencies

Map series are produced by a number of other official bodies, although the actual map printing is frequently carried out by the Ordnance Survey. These bodies include the following:

A. Ministry of Defence – 1. Military Survey – maps at various scales.
                               2. Hydrographic Department – marine charts at various scales.
                               3. Air editions of military maps or ICAO charts for air use – various scales.
B. Department of the Environment – produce maps at various scales, chiefly for planning purposes. Thematic maps include the

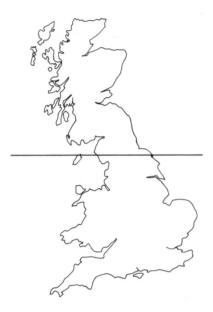

Fig. 5.1. Northern and southern sheets of the Ordnance 1/625,000 map of Great Britain.

maps of the Desk Atlas of Great Britain (about ½ million). A number of maps at a scale of 1/625,000 covering Great Britain in two sheets have been prepared by the Ministry, the Ordnance Survey and other bodies and form a series practically amounting to a national atlas in their coverage (fig. 5.1).

C. Ministry of Agriculture, Fisheries and Food. Agricultural land classification maps, 1/63,360 and smaller scales.

D. Soil Survey of England and Wales.
   Maps in a series designed to cover the whole country at the scale 1/63,360 together with special sheets, e.g. 1/25,000 Romney Marsh.

E. Institute of Geological Sciences – Geological Survey. Maps at various scales from 1/25,000 for special areas to 25 miles to the inch. The major series is the 1/63,360 'New Series' covering Great Britain on the sheet lines of the Ordnance Survey Third edition (Small sheet series). Older maps showed geological outcrop boundaries in black and were hand coloured, current series maps are colour printed and the National Grid Reference System is overprinted on the Third edition base. New revisions use a more recent topographic base, e.g. Seventh series Ordnance map.

F. Maps of overseas territories are produced by the Department of Overseas Surveys and include topographic series, town plans and general maps for many former colonies and other countries in the Commonwealth.

Land Utilisation maps

Maps of land use in Great Britain have been produced in the past on the framework of the one inch to the mile map and are currently on the sheet lines of the Ordnance Survey 1/25,000 (Second edition) covering a ground area of 10 kms north–south by 20 kms west–east. These are based on field surveys broadly carried out in the early 1960's on the six inch to the mile map and are published by the Second Land Utilisation Survey of Britain (Miss A. Coleman, King's College London).

Commercial mapping

Few topographic map series are prepared by commercial and private organisations – the chief exception being a half inch to the mile map series produced by John Bartholomew & Sons Ltd, of Edinburgh. A variety of atlases are, however, produced by such bodies, e.g. Times Atlas (printed by Bartholomews), Readers Digest Association, Clarendon Press (Oxford Economic Atlases), Wm. Collins, Sons and Co Ltd, (Collins – Longmans atlases), Thomas Nelson & Sons Ltd.

Road maps, wall maps, etc., as well as atlases are produced by other organisations, e.g. Geographia Ltd (road maps, thematic maps), A. W. Gattrell and Co Ltd, George Philip and Son Ltd group, the Automobile Association.

Further details of the types of publications of all these commercial organisations and of the official mapping agencies are contained in their catalogues, many of which are free on request, and a summary appears in the reports to the International Cartographic Association which are published in the International Yearbook of Cartography.

6 Some European map series

Before considering the map series of Western Europe it is necessary to define the various functions that any mapping body *may* undertake. These are:

1. Geodetic surveying – the accurate determination of surface position and of earth shape.
2. Topographic surveying – the detailed surveying of the topography of the country – by ground and air methods.
3. Making cadastral records – maps, deeds and ownership records, land registration.
4. The compilation of thematic maps of various types, e.g. ICAO charts, hydrographic charts, geological maps, land use maps.
5. Map production – the final preparation and printing of maps of all kinds but especially of topographic series.

The British Ordnance Survey does not itself carry out all these functions (thus, although it makes cadastral maps, records of land ownership are the function of the Land Registry office). In Western Europe the above activities are even more dispersed amongst both national and even municipal organisations as well as commercial groups. Nevertheless, close co-operation exists between the various sections of mapping and surveying since each depends on the work of others. However, each national surveying network uses an independent datum level for height calculations (fig. 6.1).

Survey organisation is frequently the responsibility of, or is associated with, a military survey body (as has been the case with the Ordnance Survey). Cadastral surveys and land registration involve the preparation of maps at a large scale (e.g. 1/500). Such scale maps are also required for planning purposes but are not necessarily needed in a printed and published form. In a number of cases this need can be met by having map originals on a stable film and making *diazo* copies as required (chapter 13). Alternatively, various local authorities may produce their own maps under special arrangements or licence.

The general topographic map needs of the military authorities may be served by the production of only small and medium scale map series. Larger scale map series may then be produced by a different national organisation or even by the major municipalities.

Country	Datum	Differences (metres)
Belgium	Brussels 1892 Ostend	+2, 33 = (Netherlands) +2, 31 = (Germany) +2, 31 = (Netherlands) +2, 29 = (Germany)
Denmark	Erritso	+0, 09 = (Germany)
Finland	Helsinki	+0, 08 = (Sweden)
France	Marseilles	+0, 25 = (Germany) +0, 19 = (Spain)
Germany (W)	Berlin	−0, 06 = (Switzerland) Denmark (see above) Belgium (see above)
Great Britain	Ordnance Datum	−0, 05 = (France)
Italy	Genoa	+0, 25 = (Switzerland)
Netherlands	Amsterdam	Belgium (see above)
Norway	Tredge	+0, 41 = Sweden
Portugal	Cascais	−0, 21 = Spain
Sweden	Stockholm	Norway (see above)
Switzerland		Italy (see above) Germany (see above)
Spain	Alicante	Portugal (see above) France (see above)

Fig. 6.1. Table showing relationships between datum levels of some European national mapping organisations (in metres) (from Imhof 1965).

In general, the main map scale at which there is considerable coverage of the countries of Western Europe and which is most readily available in a published form is 1/25,000, or 1/20,000. The major mapping bodies and map scales, etc., of the following countries will be briefly considered in the remainder of this chapter:

Belgium, Denmark, France, Luxembourg, the Netherlands, Norway, Sweden, Switzerland, West Germany.

Belgium

1. The official mapping body is the *Institut Géographique Militaire de Belgique* under the control of the Ministry of Defence. Geodetic and survey work are carried out by the *I.G.M.*, leading to the publication of topographic map series at scales from 1/10,000 to 1/250,000. Thematic maps forming part of the national atlas, soil maps of Europe and Africa are also produced by this organisation, and also administrative maps of Belgium. The basic topographic map of Belgium is at the 1/25,000 scale, with a 1/10,000 map available for some areas. Rather generalised map series at 1/50,000 (*Type Rapide*), 1/100,000 and 1/250,000 scales are also available.[1] The 1/25,000 map is a six-coloured map with a multilingual legend (French, Flemish, English) and it portrays some aspects of rural land use as well as topographic detail, contours being at $2\frac{1}{2}$ m or 5 m intervals, depending on degree of relief.
2. Large scale mapping (1/10,000 and larger) of urban agglomerations in Belgium is the role of the *Service de Topographie et de Photogrammetrie* of the *Ministère des Travaux Publiques*. The city regions of Antwerp and Brussels are so covered.
3. Thematic map series in course of production in Belgium are a soil survey (1/20,000 scale) and a geomorphological map (1/25,000 scale).

Denmark

1. The Danish official mapping agency is the Royal Danish Institute of Geodesy (*Kongelik Dansk Geodatisk Institut*). National map series are published at scales from 1/10,000 and 1/20,000 down to 1/1 million – the latter being in the style of the International Map of the World (p. 67) and longitude is referred to the International meridian and to the Copenhagen meridian.
2. Larger scale maps are prepared by the Danish Directorate of Land Registry.
3. Thematic publications include the 1/100,000 series of geological maps produced by the Danish Geological Survey and an Atlas of Denmark published commercially by Hagerup. This is a joint production involving university geographers, the Royal Danish Geographical Society and other bodies.

France

1. The official topographic map series of France is published by the *Institut Géographique National* at scales of 1/20,000; 1/25,000; 1/50,000; 1/250,000 and 1/5 million. Also relief models in shaped

1. A new 1/50,000 topographic map series of Belgium is now in preparation.

plastic are produced by the *I.G.N.* at a variety of scales, chiefly, however, at 1/50,000.

2. Larger scale maps are produced by other organisations such as the *Service du Cadastre* – the descendant of the cadastral survey initiated by Napoleon – whose principal interests are in land subdivisions and land holdings. Planning purposes are served by the *Division des Travaux Topographiques*. Both these organisations prepare plans at scales ranging from 1/500 to 1/5,000 although some smaller scale maps (down to 1/20,000) are also produced.

3. A large number of other mapping organisations exist in France of which the *Bureau de Recherches Géologiques et Minières* is probably the largest map producer of topographic maps. Marine charts and aeronautical charts are respectively the responsibilities of the *Service Hydrographique de la Marine* and the *Service de l'Information Aéronautique*.

   Studies of vegetation distribution, forestry and ecology are also served by published maps together with some detailed maps of geomorphology.

4. Commercial mapping in France includes the maps produced by the *Manufacture Française des Pneumatiques Michelin* – the well established 'Michelin' maps produced for tourism at scales varying from 1/50,000 to 1/4 million. These maps are not confined to France alone: countries covered by some 80 maps (and guide books) are France, Germany, Austria, the Benelux countries, Spain, Great Britain, Italy, Portugal, Switzerland and parts of Africa. Early maps carried mainly touring and road information but more recent editions of the 1/200,000 map series also show relief by spot heights and hill shading.

Luxembourg

1. Production of topographic maps of the Grand Duchy is under the control of the *Administration du Cadastre et de la Topographie*. Maps at 1/25,000 scale are printed by the *I.G.N.* (Paris) and conform to the sheet lines and style of the French series. Current developments include the replacement of this scale by maps at 1/20,000 of similar style and on the same sheet lines. The maps are in four colours and contours are at 5 m intervals. A 1/50,000 map series is also available.

2. Cadastral plans at 1/10,000 are produced by the Luxembourg Survey Office and copies can be obtained for administrative, planning or land registry purposes.

3. Geological maps at 1/25,000 form a complete series (1/50,000 for the Ardenne area) and are published by the *Service géologique de Luxembourg*.

The Netherlands

1. Map production in the Netherlands has a long history, and Dutch cartography and printing skills have resulted in many excellent mapping organisations – both governmental and commercial. Official maps are produced by the *Topographic Service* at Delft, at 1/10,000; 1/25,000; 1/50,000; 1/100,000 and 1/250,000 scales. The largest scale is only available in a grey outline but the remainder are at least six-colour productions. A comprehensive official catalogue is available and includes map extracts.

2. Other official map-producing bodies include the *Waterstaat* (Transport and Water Management), Cadastral Service, Geological and Soil Surveys, Hydrographic Bureau, etc. Water control maps, soil and geological maps are all at 1/50,000 scale. Detail maps at a larger scale are produced for specialised needs, e.g. maps of the Rivers Maas and Rhine at 1/2,000 and 1/5,000 for maintenance purposes. Smaller scale maps are also produced by the Geological Survey including a detailed map at 1/600,000 with text in Dutch and English.

3. Large scale maps (1/10,000, 1/5,000, etc.) are prepared also by the municipalities from their own as well as from the official surveys. These, together with thematic maps, planning maps, etc., are available for the major urban areas – Amsterdam, Rotterdam, the Hague.

4. Thematic maps are well represented in the sheets of the national atlas (basic scale 1/600,000) – *Atlas van Nederland* and the historical and agricultural atlases of the Netherlands. Commercial map production shows a specialisation in reproduction of old maps and atlases, e.g. *Theatrum Orbis Terrarum*, reproducing atlases of the sixteenth, seventeenth and eighteenth centuries. Road maps are produced by the Royal Dutch Touring Club, the *Cartografisch Instituut*, Bootsma; atlases by *Wolters – Noordhof, N.V.* and the *Elsevier* Publishing Company.

Norway

1. Maps at scales of 1/25,000 and smaller are produced by the official organisation – *Norges Geografiske Oppmåling* – one of the oldest European bodies of this kind, founded in 1773. The basic map scale is the 1/50,000 and longitudinal references are to the Oslo Meridian as well as to the international graticule.

2. Geological maps form a complete series at 1/250,000 scale and hydrographic surveys of home and overseas waters are made at various scales by other official organisations. Polar areas and Spitzbergen are mapped by the *Norsk Polarinstitutt.*

3. Large scale survey maps are prepared by some of the municipalities for planning and other purposes and a commercial map producer is *Cappelen* who publishes tourist and road maps at small scales (e.g. 1/325,000 and 1/400,000).

Sweden

1. There are a number of important map-producing agencies in Sweden. The official Geographical Survey office (*Rikets allmänna kartwerk*) produces the topographic map on the basis of a Transverse Mercator grid (central meridian 2° 15′ W of Stockholm) but the actual printing is carried out by either private or state printers on a contract basis. Map series include a photomap and the Economic Map of Sweden (at 1/10,000 and 1/20,000 scale) as well as the 1/50,000 topographic series (1/100,000 in mountainous areas). The Economic Map is based on the large scale photomap which it has generally superseded and is a four-colour orthophotomap in which the background tones of the rectified air photographs are printed in different colours according to land use, e.g. yellow for arable land, planimetric detail in black and a green tone elsewhere. Contours are overprinted in brown.
2. Land survey requirements in Sweden are undertaken by the *Kungliga Lantmateristyrelsen* (National Land Survey Board). Property divisions are normally recorded on black editions of the 1/10,000 base map but larger scales (1/400 and 1/1,000) are also used. This organisation is also involved in the compilation of a new Property Register in which map grid co-ordinates will be used to identify plots. This register will form part of the Swedish National Data Bank (Hägerstrand 1969).
3. Other official bodies are concerned with the production of marine charts, aeronautical charts, geological maps, etc., and another organisation handles sales and distribution of official maps.
4. A large commercial organisation – the *Esselte Aktiebolag* or Esselte map service – is one of the largest map-printing bodies and produces many of the official map series for the Geographical Survey office as well as tourist maps, atlases, town plans, etc., for its associated publishing houses.

Switzerland

1. The national topographic maps of Switzerland are produced by the official organisation, the *Service Topographique féderale*, at scales of 1/25,000, 1/50,000 and 1/100,000.
2. Larger scale maps (1/5,000 and 1/10,000) are the responsibility of the *Direction fedérale des Mensurations*. These form the basis of maps for planning and land registration requirements.

3. Thematic maps and maps of special areas are well represented in Swiss cartography, ranging from geological and vegetational small scale surveys to a large scale (1/10,000) map series of the Aletsch glacier. This latter shows many of the features common to Swiss map series, being prepared from aerial and ground surveys, with extremely fine colouring and printing details of the complex structure of this area. Contours are shown at 10 m intervals, brown or black to indicate relief shading on rocky areas and blue on the ice areas. Rock outcrops are drawn and lettering is very neatly executed. Hill shading techniques, the use of pastel colours and light tones, have been developed by Swiss cartographers over many years to a very high standard.

4. Commercial mapping organisations are well represented in Switzerland, the firms of *Kummerly & Frey, S.A.*, and *Hallwag, S.A.*, being well known throughout most of Europe for motoring maps as well as thematic maps, town plans, etc.

5. The Swiss *Kartographisches Institut* at the Federal Polytechnic in Zürich has also been responsible for a number of cartographic productions and research in the subject. Probably the most widely known of its sponsored publications is the *Atlas of Switzerland*, edited by Professor E. Imhof, which uses the 1/500,000 map of Switzerland as its main base map.

West Germany

The main survey network for the Federal Republic of Germany is under the direction of a federal organisation, the *Institut für Augewrandte Geodäsie*. This organisation also produces smaller scale maps of the country (scales of 1/200,000 and smaller). Other federal organisations are the *Deutsches Hydrographisches Institut* (soil survey), *Bundesamstalt für Bodenforschung* and ministries concerned with federal planning and statistics. Again these are chiefly responsible for small scale publications or atlases.

Larger scale topographic map series and also thematic maps depicting soils, geology, etc., are the work of the various states (Länder). Although prepared and published by the Länder, these maps conform to an overall sheetline system and specifications so that there is available full topographic coverage for the country. Certain areas are only available in older monochrome styles at the largest scale (1/25,000) but generally this is a fully coloured map depicting contours, hill shading, communications, buildings, vegetation, etc., in conventional colours of brown, blue, black, red and green.

Much progress remains to be made with larger map scales but recent work on an orthophotomap at 1/5,000 (of Nordrhein-Westfalen) (fig. 4.2) may enable gaps to be filled in extensively built-up areas. Contours on this map are shown by black overprinted lines, names and roads in white.

Commercial mapping organisations in West Germany show a strong concentration on atlas production, e.g. *Kartographisches Institut Bertilsmann, Georg Westermann Verlag, Verlag Herder KG*. Road maps are produced by the *Kurt Mair* organisation and *Falk-Verlag of* Hamburg.

7 Atlases and world map series

An atlas is defined as 'a collection of maps designed to be kept (bound or loose) in a volume' (*Glossary of Technical Terms in Cartography*, 1966). The term 'atlas' generally invokes a somewhat more limited image – that of the fully bound reference volume containing maps of most countries of the world. On the other hand, many specialised atlases have been produced depicting features of a limited area. Atlases can perhaps be conveniently classified as

1. Library and reference atlases,
2. School atlases,
3. National atlases,
4. Regional atlases.

Common features of the first two categories are that they include collections of maps of both topographic and thematic types covering different countries, regions, continents and the world. The amount of detail shown on such maps depends on the purpose of the atlas, e.g. junior school, secondary school or the gazetteer and reference function of the library atlas. The selection of areas covered in the atlas also reflects the purpose of the publication. The 'concentric' approach may be followed in school atlases with special neighbourhood maps, aerial photographs, etc., of the local region forming the introductory section. This will be followed by comprehensive topographic and thematic map coverage of the home-land and less detailed maps covering a wider area such as the continental area and the rest of the world.

Atlases produced in any particular country tend to contain map coverage of that country to a greater depth than 'foreign' areas. This may be obtained by including maps at a larger scale or by selecting and stressing particular detail. Nevertheless the common feature of both scholastic and library atlases is their firm binding (as opposed to a loose-leaf format) and their inclusion of an index or gazetteer. This generally uses the conventional system of geographical co-ordinates of latitude and longitude but other methods are also used, for example, separate grid systems of numbers and letters for each page, or the system of time zones (cf. *Bartholomews Advanced Atlas*).

As a form of national inventory the atlas of an individual country is a valuable document but may sometimes have a prestigious aim. Since a full and adequate documentation in map form takes time to

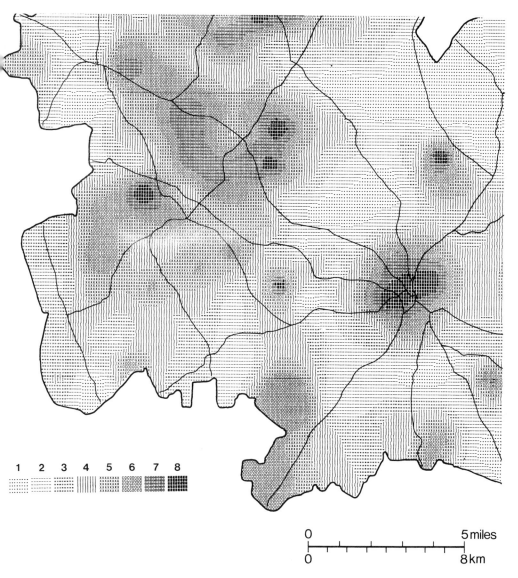

Fig. 7.1. *Computer map in regional atlas.* (*From* Character of a Conurbation, *Rosing and Wood 1971. Map 14*). *The percentage of dwellings that are rented by local authorities. Shading obtained by overprinting symbols to give a range of eight equal stages to cover 100%.*

prepare and individual map sheets can quickly become outdated or need revision or replacement, it is normal to produce national atlases in a loose leaf form. These may then be mounted in a folder or contained in a wallet or box.

National atlases usually aim to present all possible distributional data for the country concerned and can be very extensive in their coverage. An introductory section will probably include maps of such features as relief, drainage, hydrographic regimes, soil types, climatic records, etc. Resources and economic data – industrial and agricultural production, employment, trade, communications and similar information will form another section. Demographic facts usually yield a further set of maps and special characteristics of the country may give rise to specialised maps, e.g. water and flood control features in the Netherlands, avalanche hazards in Alpine regions, vegetation patterns in tropical countries.

The preparation of regional atlases has followed the collection of data needed for planning purposes. In many respects the maps produced for these purposes will have much in common with national atlases and background information, historical, topographic and other details will provide an introductory set of maps. Other sections will reflect both the region and the purpose for which the atlas has been produced which may be purely for reference purposes or it may be as a basis for planning or, again, to set out planning proposals. The tables of contents of the Desk Planning Atlas of England and Wales and the Atlas of the Netherlands are given in Appendix IV. The scope of a planning portfolio is indicated by the former and that of a national atlas by the latter. Within these main groups of atlases large variations occur in the types of map. The common factor is the presence of at least some topographic maps but by far the greatest number of maps in atlases of the national and regional type will be of the thematic form, both to illustrate distributions and to present plans and concepts. Large scale maps and plans will probably make up a major part of planning atlases with medium scale maps characterising regional atlases and smaller scales used for national atlases. If a considerable amount of relevant and up to date data is to be presented this can be done by using methods of automated cartography, e.g. computer printouts, as in the Birmingham and West Midlands atlas (fig. 7.1).

International Map Series

Although international co-operation on a proposed world map series began with its proposal at the 1 to 1 million scale at the 1891 meeting of the *International Geographical Congress*, progress was inevitably hampered by the wars and economic disturbances of the succeeding decades. Nevertheless, the map series was launched at the Paris

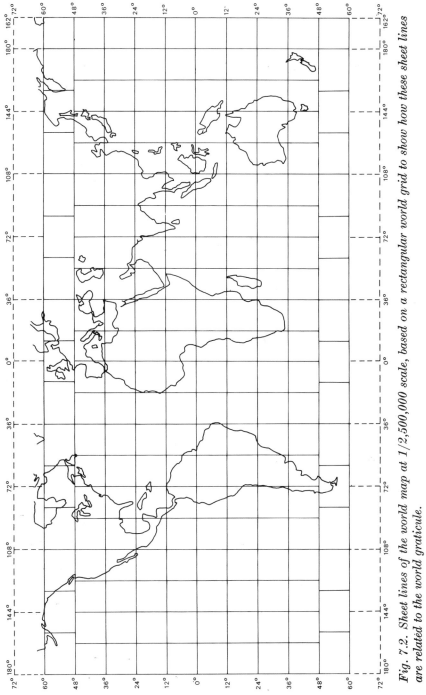

Fig. 7.2. Sheet lines of the world map at 1/2,500,000 scale, based on a rectangular world grid to show how these sheet lines are related to the world graticule.

Conference in 1913 when decisions were made as to the projection – a modified Polyconic for most of the earth's surface – and the methods of depicting relief, hydrology, populous areas and routeways. Work continued by the various member countries of the *International Geographical Union* and by 1949 approximately one-fifth of the map sheets had been published (405 out of a total in excess of 2,000). About half of these were to the full specifications of the *International Map of the World* (*IMW*), others being compiled by member countries with the same pattern, style and sheet-lines. Since 1954 the General Secretariat of the IMW has been part of the Cartographic Office of the United Nations and discussions have also continued as to the comparative roles of the IMW and the ICAO Aeronautical Chart at the same scale.

In 1962, at the Bonn Conference on the IMW, alterations were made to the specifications and sheet-lines, and the general objectives of the map were agreed as follows:

(a) to provide, by means of a general purpose map, a document which enables a comprehensive study of the world to be made for pre-investment survey and economic planning and also to satisfy the diverse needs of specialists in many sciences.
(b) to provide a base map from which sets of thematic maps can be prepared (e.g. population, geology, vegetation, administrative limits, statistical evaluation).

It must be noted that vegetation cover is not included on the base map but particular encouragement was to be made to complete such maps for the developing countries.

The ICAO World Aeronautical Chart at the 1/1 million scale was developed out of a United States Air Force series at this scale and its completion has been given priority by many individual countries.

A further map series at the 1/2·5 million scale initiated by the national cartographic office of Hungary should also be mentioned. This began publication in 1964 and will eventually cover the world in 244 sheets (fig. 7.2). The official languages of the map are Russian and English, and the purpose of the map series is to meet the need for a world base map of a general geographical nature (Robinson 1965).

At the larger scale of 1/250,000, a further map series is also under way. This is to have two versions – a Ground (or Land Map) and an Aeronautical (or Air Chart) form (Meine 1966).

8 Cartographic appreciation

Maps are frequently so much an integral part of modern existence that they are taken for granted: the commonplace map in a familiar style becomes the norm and its value remains unquestioned. Faced with a new map type the user is tempted to make quick decisions of a subjective kind or may indeed use the map in an inaccurate and inadequate way. It is therefore necessary to be able to assess the value of any given map in a fair and as objective a manner as possible. This assessment is variously regarded as 'appreciation' or 'critical appraisal' or 'critique' of the map. The chief dangers in making such an evaluation are those of subjectivity and of an intolerance bred from years of habitual use of maps of a limited range of types. The suggestions in this short chapter fall into both objective and subjective assessments which can be made of any individual map or series of maps.

1. *Objective assessments.* Such observations must necessarily be confined to dealing with precise details of the map under examination, namely

    (a) the overall features of the map, and
    (b) methods utilised for presenting the mapped data.

2. *Subjective assessments.* Under this heading come all those factors in which personal choice or preference enters, even if only to a limited extent. Such factors fall into the following groups:

    (c) suitability of mapping techniques;
    (d) adequacy of detail depicted for map purpose;
    (e) aesthetic considerations.

Considering these points in turn:

(a) *The overall features of the map*
These include:
The size and format of the published map sheet(s) – including presence and size of margins.
The layout of the printed sheet (fig. 8.1) –
    title
    sheet number
    legend

scale(s)
additional data
date of survey
map production
reliability diagram
grid and its explanation

At this stage in examining the map the assessor should merely record these various points in a purely quantitative form, i.e., size of the map in appropriate units, and the position of the other details listed above, noting their presence or absence.

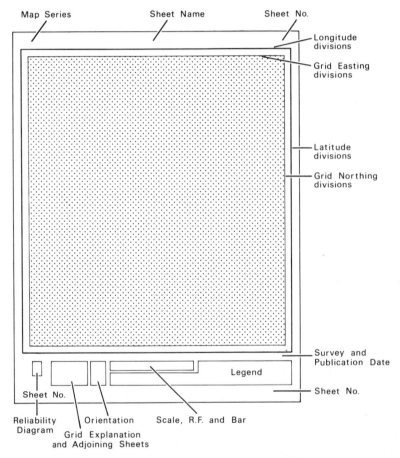

Fig. 8.1. Annotated diagram to show layout of typical Ordnance Survey one inch to the mile map and position of legend, scale, marginal information, grid and graticule numbering. Shaded area is that occupied by actual map information, outer line shows dimension of sheet of paper on which the map is printed.

(b) *Methods utilised on the map under consideration*

Once again it is necessary to work through the various items shown on the map and to note the various methods used, which, if they are not standard or conventionally accepted methods, should be referenced adequately in the legend.

Thus it is essential to consider, on a topographic map, how the following features are indicated:

Relief (contours, spot heights, shading, colour, other methods).

Drainage (distinctions between different forms, widths, etc., flow, water sources).

Coastlines (type, tide marks, foreshore).

Hydrographic features (symbols, contours).

Vegetation (shading, symbols, boundaries).

Land subdivisions (plot boundaries, fences, etc.).

Buildings and apparent generalisation (will be indicated when considered in relation to boundaries which may, however, themselves be generalised).

Settlements and their naming.

Communications (road, railway, canals, transmission lines).

Administrative divisions – boundary, symbols and naming.

Thematic data – isopleths or choropleths.

The adequacy or otherwise of the legend can be considered here and omissions, errors or departures from accepted practice as these are clearly objective observations.

(c) *Suitability of mapping techniques*

It is frequently difficult to assess this point without some knowledge of either the type of landscape depicted on the map, or without having attempted some precise investigation from the map. The latter course is always to be recommended because the map is a tool to be used and without having experience of the tool its value must remain in doubt. For example, a topographic map can be examined with a view to isolating the major relief units or the components of the transport network. Similarly a thematic map should be studied in order to extract information relating to a random choice of points. The ease of these various operations will provide a clear measure of the map's suitability (and also the adequacy of the legend or other explanatory text). The degree of subjectivity of the map assessment can thus be somewhat lessened. Nevertheless, much of this part of the exercise will necessarily be confined to an examination of the suitability of methods under (b).

(d) *Adequacy of detail depicted for map's purpose*

Whilst studying the mapping techniques used it is quite likely that this next point of adequacy will also have been considered. The essential point, however, is the *purpose* of the map. The suitability of the techniques (above) will merely indicate whether the map can be interpreted as a model of the landscape. Under this heading the whole validity and *raison d'être* of the map must now be examined. It is often pertinent at this stage to examine also the relations of the map sheets to one another and to look again at the objective points recorded earlier, e.g. map size, layout, additional or marginal data.

(e) *Aesthetic considerations*

This section must embrace a wide range of topics and although the aesthetic and artistic features of the map should be considered there are other related points to be examined first. The general handling ability of the map (which stems from overall layout, suitability, etc.) and its overall suitability having been studied it remains to be seen if the presentation is attractive. Many maps are prepared on a commercial basis and the overall impression which they give to the user is a vital selling point. They must also contain the information the user wants and be shown to contain this information. Methods of folding, covering and general salesmanship are not unimportant here.

Part III

Using maps

Fig. 9.1. To illustrate the fact that any horizontal projection will cover a smaller area than the true ground surface.

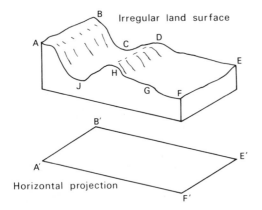

Fig. 9.2. Shape (a) shows the precise shape of a contour which becomes generalised to shape (b) for reproduction at a smaller scale. This generalisation tends to be greater in area than the original.

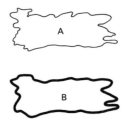

9 Measurements from maps

This chapter will consider the general problem of extracting quantitative data from maps, it is concerned with various precise measurements that can be made and also how these measurements may be used in attempting map analyses of the landscape.

Detailed topographical information is frequently required by geographers, planners, etc., but it must be remembered that errors can occur whilst the data is being obtained from the map.

These errors include:

(a) *Scale errors*. These can be due to variations in the size of the paper and can be reduced by always referring map measurements to the printed scale bars on the surface of the maps. These will change their shape in the same way as the features on the map itself. Such errors can be either positive or negative.

(b) *Projection errors and curvature of earth's surface*. On small scale maps errors are introduced in representing the curved surface of the earth on a flat sheet of paper. Errors due to the type of map projection used are negligible on medium and large scale maps. They can be calculated by referring to the type of projection graticule used. (See standard textbooks on map projections.)

(c) *Slope and height errors*. All information on the map is presented as if on a horizontal plane, usually that of mean sea level: it is not presented as on the natural undulating surface of the landscape. If extreme accuracy is required allowance must be made for the difference between these two surfaces or alternatively it must be made clear if distance, areas, etc., refer to the complex natural surface or to the more usually accepted sea level plane. These errors are such that the mapped information is smaller than the true value (fig. 9.1).

(d) *Generalisation errors*. As has been indicated in chapter 1, some degree of generalisation is required between the details of the landscape and its representation on the map (fig. 9.2). Such generalisation has the effect of smoothing out irregularities. Under this heading can be included errors in the original draughting of the map. Distances measured may therefore be somewhat too small compared with ground distances but on the other hand, areas can be too great. A check can be made, by summing areas to compare with a known value, such as that of an administrative area.

(e) *Measurement or instrumental errors*. Such are either gross errors due to carelessness, mistakes, etc., or systematic errors resulting from the method used. In all cases measurements to printed lines on the map should be made to the centre of the line but many discrepancies can occur, e.g. when measurements are made to a printed road symbol – which may well be considerably enlarged from its true dimension – these might be taken to the centre of the colourfill or they may be made to the appropriate black lines on either side of the colour. Where a measuring instrument is being used such as the planimeter for areas (see below) then a recognised 'drill' must be followed to reduce errors in the instrument.

It must be stressed that all measurements made from maps are subject to errors of the above nature and that although allowance can be made for them, it is meaningless to perform further calculations with these measurements to levels beyond their accuracy or significance.

Types of measurement

The main types of measurement made from maps are of distance, direction and area. *Distance* measurement clearly involves assessing lengths on the surface of the map – either against each other for comparison purposes – or against the known scale to obtain absolute values. Direct or straight line distances may be required and present little difficulty as they can be taken directly with a pair of dividers or a scale placed on the map surface. More usually, distances are required along erratic paths: sometimes these can be broken down into sections and each part taken with dividers or it may be necessary to follow the active path with a map measuring wheel (*opisometer*). Any splitting of the distance into parts will introduce possible cumulative errors, either small and systematic or large and accidental (Maling 1968).

The measurement of *direction* is less often required but produces more problems than either distance or area calculations. The problem here is that of datum direction – usually that of north. Well produced maps

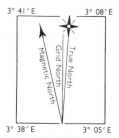

Fig. 9.3. Some typical North points (grid, true and magnetic) from an Ordnance Survey one inch map, with accompanying notes to indicate more exact differences between these directions.

True North
Difference from Grid North at sheet corners is shown above

Magnetic North
About 8° W of Grid North in 1970 decreasing by about $\frac{1}{2}$° in eight years

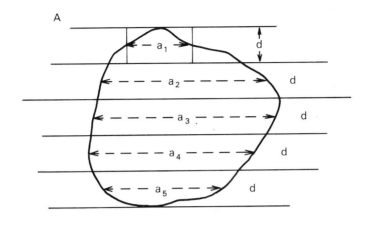

A

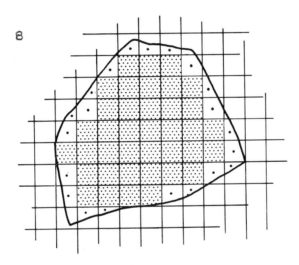

B

Fig. 9.4. Area measurement methods.
A. A series of parallel lines, equally spaced at distance d, cross the area to be measured. The distance between the shape's outer boundaries midway between each line is then measured $(a_1, a_2, a_3,$ etc.). Each strip can then be considered as a rectangle of area $d.a_1, d.a_2,$ etc., and these are summed. Alternatively, vertical lines are drawn on the 'give-and-take' principle, to omit an estimated section equal to an outer area which is included in each rectangle. The area is still given by $d(a_1 + a_2 + a_3 \dots)$.
B. The area to be measured is covered by a regular pattern of squares (e.g. sheet of graph paper placed beneath the map on an illuminated tracing table or the area is traced onto squared paper). The total number of whole squares is summed, together with all partial squares. Area = Sum of whole squares + Half total of part squares.

will generally indicate at least the direction of true north. Topographic sheets will also add that of magnetic north – for the benefit of field map-users and probably a note as to the relationship between grid north and true north – a relationship which will vary from sheet to sheet (fig. 9.3).

Area measurement is usually carried out on the surface of the map and assumes that the discrepancy between this sea level plane and the undulating natural surface is smaller than the errors of making the measurements. There are numerous ways in which areas can be calculated (fig. 9.4) and summaries appear in most standard textbooks. Frolov and Maling (1969) provide a discussion of the relative accuracies of these methods. The most usually available mechanical method is that of the fixed area polar *planimeter*. In this instrument a measure is made of the movement of a rod whose locus is constrained by having one end fixed to a radial arc (fig. 9.5).

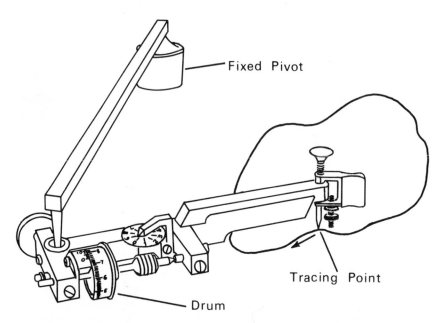

Fixed Pivot

Tracing Point

Drum

Fig. 9.5. The fixed arm polar planimeter. This measures area by tracing the outline of the closed area under consideration in a clockwise direction and returning exactly to the starting point. The movement of a wheel at right angles to the planimeter arm is recorded in terms of whole revolutions and decimals of a revolution. The area swept out is given by this figure multiplied by the constant of the instrument. Since very small movements are measured, a vernier reading of the drum is necessary and it is also necessary to repeat the measurement, say, three times and to take the mean of these three provided they show reasonable agreement.

The area to be measured is traced along its perimeter in a clockwise direction with an index mark; starting from one convenient point to which the index must exactly return. Reading of the recording dials before and after so tracing the area's perimeter will give a value in instrumental units. These are in terms of complete and decimal fractions of revolutions of the wheel attached to the planimeter rod. They must be multiplied by some constant for the particular instrument to convert into areas in square inches or square centimetres. In order to obviate gross errors, at least three readings of each area should be taken and the mean calculated, provided there is reasonable agreement between the three readings. Values showing wide disagreement should be disregarded and the readings taken again.

Non-instrumental methods of measuring area are also possible. Of these the most simple is probably that of transferring the required area to a grid of printed squares either by tracing off or by superimposing transparent printed tracing paper (e.g. with 1/10 in squares or 2 mm grid). The number of whole squares (W) and partial squares (P) are separately summed and the area is approximately given by

$$A = W + \frac{P}{2}$$

i.e. the total of all complete squares plus half the total of partial squares. Alternatively a visual assessment may be made so that areas excluded in some partially occupied squares are balanced by areas included elsewhere (fig. 9.4).

This leads to another method of area measurement in which the area is covered by a series of parallel, equally spaced lines (fig. 9.4). Bounding lines at each end of the parallels can be drawn on this 'give-and-take' method but almost certainly an irregular shape will remain as in fig. 9.6. This is then measured by other methods, e.g. Simpson's Rule (below). Geometrical calculations of area (mensuration methods) provide an absolute value and are possible in cases where regular outlines are considered – rectangles, parallelograms, triangles, sectors of circles, etc.

Where it is possible to reduce the shape to an irregular outline on one side and a straight line on the other, standard mathematical methods of area calculation can be used – either Simpson's Rule or the Trapezoidal Rule (fig. 9.6). This will require the measurement of a series of regularly spaced ordinates at right angles to the straight side.

Area measurements involving a 'sampling' technique may also be used: these have been described by Haggett (1965) and Frolov and Maling (1969) and have been used in conjunction with the Second Land Utilisation Survey of Britain. They yield a high accuracy, coupled with speed of operation, and require no measurement of areas

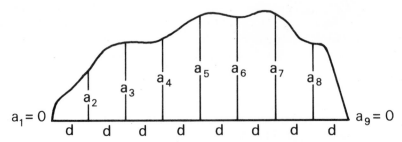

Fig. 9.6. Area calculation by ordinate measurement. If a regular series of vertical ordinates are measured to the irregular boundary of the area from a straight line either of two formulae may be used in order to calculate the area:
(a) *Simpson's rule*

$$Area = \frac{d}{3}\,(a_1 + 4a_2 + 2a_3 + 4a_4 + \ldots + 2a_{n-2} + 4a_{n-1} + a_n)$$

(b) *Trapezoidal rule*

$$Area = d\left(\frac{a_1}{2} + a_2 + a_3 + \ldots + a_{n-1} + \frac{a_n}{2}\right)$$

where d is the common distance apart of the ordinates a_1, a_2, etc. Note that an ordinate must be measured at both the extreme ends of the lines. If the outline approaches the horizontal axis this ordinate will become 0 but it must be included in the formulae. With Simpson's rule it is necessary that there are an even number of divisions (i.e. the number of ordinates must be odd).

as such but only the counting of the number of occurrences of the data under examination over the study area, map sheet or series of maps. Various sampling patterns are illustrated in fig. 9.7 and these are normally marked on a transparent overlay which is then placed over the data map. A count is then made of the activity, e.g. land use, at each point and these are summed for the area or map sheet. If the area extends over several map sheets care must be taken neither to exclude the join of the sheets or to count it twice.

Angular measurements on maps to ascertain aspect, bearings or other criteria best quoted in degrees or radians must be made with caution. Large scale topographic map series, however, are generally on an orthomorphic projection and thus angular properties on the ground and the map are identical. The only problem remaining is that of a datum or starting point for bearings. True north, magnetic north and grid north should be quoted in the map margins or the relationship between some of these directions and lines printed on the map indicated. Angles and bearings on non-orthomorphic projection smaller scale maps cannot be accurately measured and only general statements made in relation to cardinal compass directions. Because angles are frequently measured on a whole circle basis, a 360° protractor should be used.

Measurement can be used in map analyses in many different forms, either to produce new diagrams which will themselves be analysed, or a statistical analysis may be made of the extracted data. (Much of the latter is beyond the scope of this volume.)

Diagrams produced from map measurements

Under this heading can be included such studies as the construction of profiles, block diagrams and summary diagrams such as network patterns, and other 'models' of the landscape based on the map evidence.

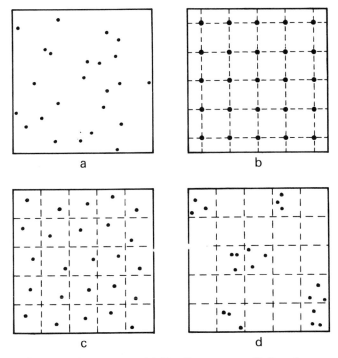

Fig. 9.7. Point sampling patterns. (a) Random pattern. (b) Regular, or systematic pattern. (c) Systematic unaligned random pattern. (d) Random pattern within specified grid squares (nested random sampling). Sampling methods may be used in order to choose locations for further study instead of undertaking a tedious study of a complete area. In area measurement usage, sampling over a whole region will consist of counting areal activities at the sample points and using the number of occurrences divided by the total number of points as an indication of the proportion that activity bears to the total area.

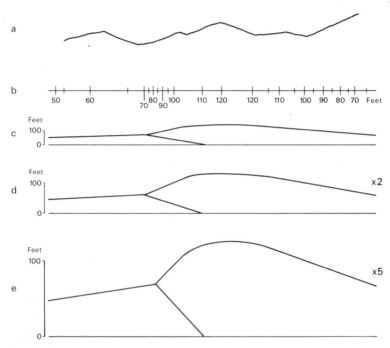

Fig. 9.8. Profiles. Simple profiles are drawn along either straight lines or, more usefully in analysis, along watersheds, river courses, or at right-angles to successive contours to show maximum slopes. The plan view may thus be as in (a) but this must be reduced to a straight line (b) before construction of the profile as a simple line graph (c). Horizontal and vertical scales need not be the same; a vertical exaggeration can help stress relief or slope changes: (d) and (e) are the same profile with ×2 and ×5 vertical exaggerations. However, if geological dip lines are inserted, shown by the sloping lines, the effect of vertical scale exaggeration will be to increase the angle of dip to a ridiculous extent.

Profiles

The method of obtaining a 'side elevation' view of the land form is shown in fig. 9.8. Spot heights and contours provide a series of discontinuous values for height which can be plotted against distance along the selected line of profile. The latter can be a straight line between two points, a grid line, or can be a sinuous line such as a watershed, road or stream line. The resulting single profile can be made to yield more clues as to the physical landscape if it is then compared with the geological map so that underlying rock information can be added and so the profile is converted into a cross-section.

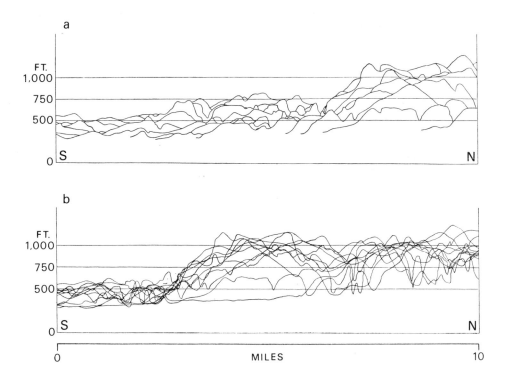

Fig. 9.9. Profiles used in preliminary morphological investigations. A series of regularly spaced parallel profiles may be combined to search for possible regional patterns as in (b), superimposed profiles. The effect of a panorama is produced by the method of projected profiles (a), in which only those sections are drawn that are not obscured by nearer profiles. These profiles are drawn along south–north grid lines on the southern edges of the Pennines (from Clayton, K. M. (1953) Denudation Chronology of the Middle Trent. Trans. Inst. Brit. Geogr. 19, 25–36).

Profiles can be combined in a variety of ways in order to establish hypotheses of landform evolution (serial profiles, projected profiles, fig. 9.9) or they can be used as a basis for block diagrams in order to present a perspective pseudo three-dimensional view of the landscape.

Nevertheless, profiles must be used with care, the most important consideration usually being that of the vertical dimension. A true to scale profile is usually rather flat in appearance and a *vertical exaggeration* is frequently applied. This immediately makes all slope angles incorrect and, in the case of geological sections, increases the apparent dip of rock strata. For some purposes a massive exaggeration may be necessary, particularly in order to draw attention to relict landform features but it is unwise to exceed 5 × horizontal distance: a common dimension is that of 1 in to 1,000 ft vertically, combined with a horizontal scale of one inch to the mile.

Block Diagrams

The need for a pseudo three-dimensional diagram is often encountered in the interpretation of landforms but is not confined to studies of the physical landforms since population and agricultural production figures for example may also be used as the 'z' component.[1] Various types of block diagram have been developed (Monkhouse and Wilkinson 1971) but the most conveniently constructed is the simple isometric diagram prepared from a series of profiles or from transformed contour patterns (figs. 9.10 and 9.11). In addition, a mechanical method of drawing block diagrams is available in the *Perspektomat* – a device which incorporates a pantograph mechanism so that not only is the appropriate viewing angle transmitted to the diagram but also changes of scale are possible.

Slopes

Although it is easy to measure slope angles from a contoured map by scaling off the distance between two successive contour lines, the student of landforms is sometimes concerned to obtain 'regional' values of slope or 'average' slope over a wide area. On some maps, notably those of France and Germany (Appendix II), the calculation of hillside slope is made easier by the inclusion of an additional scale. A number of

1. i.e. assuming the two-dimensional map is in the form of x and y directions and that the third dimension of height or other function is z.

Fig. 9.10. Block diagrams.
(a) *A perspective block diagram, in which the sides and detail will appear to converge towards two vanishing points on the perspective horizon.*
(b) *A simple isometric block diagram, one side inclined.*
(c) *An isometric block in which both sides are inclined.*
(d) *Extract of simple contour pattern covered by a regular grid as a basis for the construction of block diagrams (e) and (f).*
(e) *Construction of a block diagram by reconstruction of the contour pattern in a series of 'layers'. The detail of each of the squares in (d) is transferred to the isometric grid at the appropriate level. This is the basic method of the* Perspektomat, *an instrument which aids in the drawing of block diagrams.*
(f) *Construction of a block diagram by drawing successive profiles across the terrain (some are indicated in broken lines) and finally drawing hachures or shading down the slopes.*

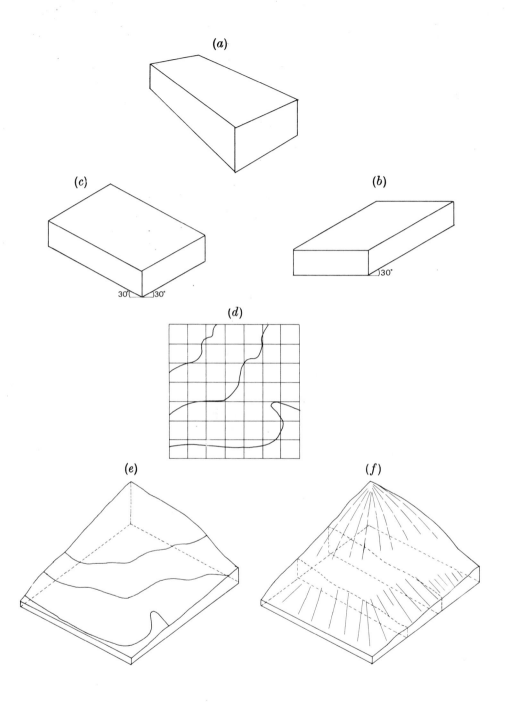

(a)

(c)

(b)

)30°

30° 30°

(d)

(e)

(f)

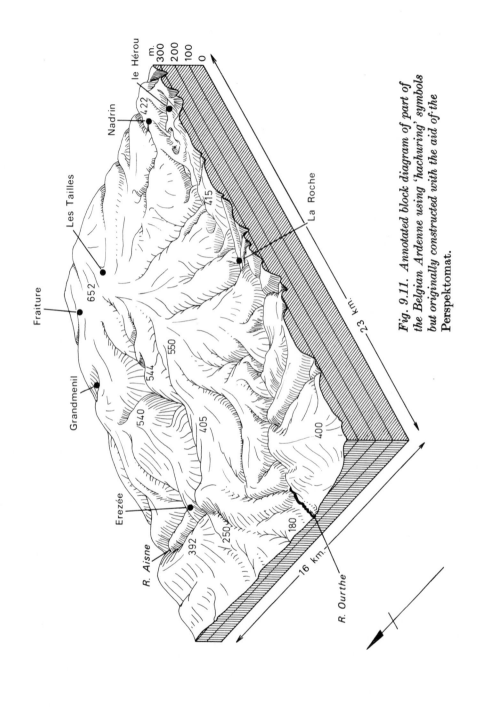

Fig. 9.11. Annotated block diagram of part of the Belgian Ardenne using 'hachuring' symbols but originally constructed with the aid of the Perspektomat.

complex methods have been devised for estimating average slope and these are described in Monkhouse and Wilkinson (1971).

Other diagrams

Much of the data extracted from measurements will result in tables of data. This may then need representation in a cartographic form, or more normally, by some statistical diagram – a graph of some kind (see pp. 90–1). Simple linear diagrams to indicate networks, hierarchies or other relationships may also be drawn and dimensioned by writing the appropriate values, such as distances, on the diagram (fig. 9.12).

Data relating to individual points can be presented in a map form merely by indicating the point (e.g. by a dot or cross symbol) and entering the value of the data alongside (fig. 9.12).

It is possible, however, that values for the data can be expected to occur for positions for which no precise reading is available. In other words, data values must be interpolated, provided it is known that the phenomena so recorded has a continuing occurrence over the whole area. If the phenomena is known to be of isolated and separate occurrences then no continuum can exist and any interpolation is purely hypothetical.

The interpolation of values between known values forms a way in which isolines can be constructed; or, conversely, from a map of isolines it is possible to suggest the value of a particular phenomenon at any given point. Interpolation or extrapolation of data from simple graphical figures requires an understanding of the way in which the graph is behaving – or the relationship between the graphical data. This relationship can be approximated within statistically calculated limits known as 'confidence limits' and the behaviour of the graph or functional relationship between the variables is expressed by the 'regression'. In the case of interpolation of isolines or the extraction of data from isoline maps it must be assumed that there is a constant change of data between the isolines or in all directions from a given value (fig. 3.8).

It is possible for many of the operations referred to above to be carried out by a computer program, providing that the initial map data is already available in co-ordinate form (i.e. Cartesian co-ordinates and value of data). The tedious calculation of rates of change between adjacent points are then performed by the computer and the position of the isolines obtained.

More sophisticated analyses of data can be made by extensions of the methods of interpolation and/or calculations of regression values mentioned above. Such analyses may be required in order to examine the three-dimensional properties of mapped data and their functional

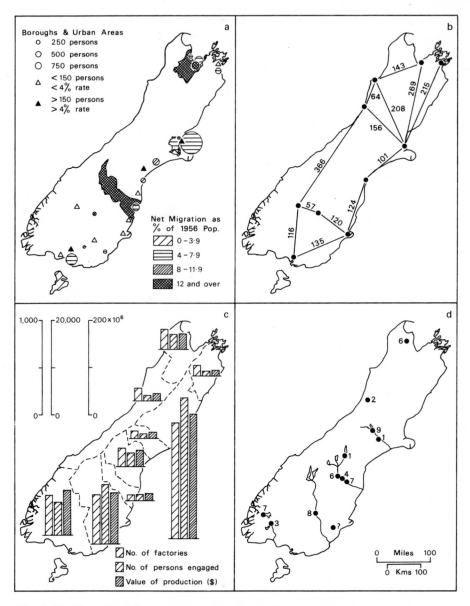

Fig. 9.12. Numerical data on maps. New Zealand, South Island, with statistics illustrating various aspects of human geography.

(a) *Population gains by migration 1956–1961.*
(b) *Distance in miles between main centres.*
(c) *Factory production 1967–1968 (all industries).*
(d) *Electric power stations, number of sets.*

relationships in three dimensions. In these cases the techniques of study are referred to as a study of 'trend surfaces'. The surface in question is the solid form of a mathematical relationship in 3 dimensions (x, y, z) which is the graphical expression of progressively more complex functions – quadratic, cubic, etc. (fig. 10.7). The degree of approximation to a selected surface can be tested and divergences from the surface (or 'residuals') can be calculated.

LAND   USE

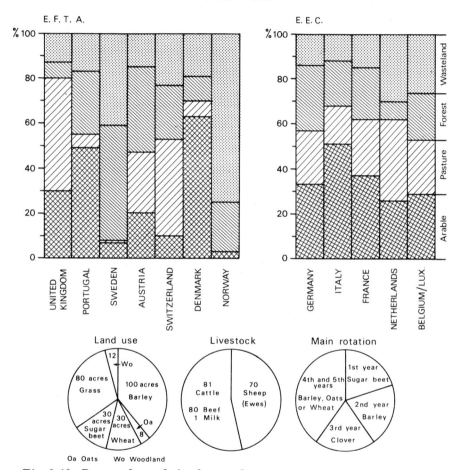

Oa Oats   Wo Woodland

Fig. 9.13. Rectangular and circular graphs.
(*Above*) *Each rectangle is divided according to the percentage area in each land use classification shown on the right. EFTA countries on the left and EEC countries on the right. It would have been possible to make each rectangle proportional to the total area of the country by varying the widths.*
(*Below*) *Circular graphs divided according to percentages of various distributions, areas or occurrences. 360° is equal to 100% and the circle is divided accordingly.*

10 Map analysis

In the previous chapter attention was paid to ways in which precise information may be extracted from various types of map. The map may be regarded as a representation or *model* of the landscape and is a major source of material for geographical studies, i.e. a data bank; it has, however, been prepared for some specific purpose and may well be selective or even biased in this representation.

Much modern work in geography and allied subjects is in the field of analysis – locational, spatial, regional, etc. Map analysis may provide a starting point for some of these investigations but it generally needs to be supplemented by other observations, reports, census data, etc.

Geographical investigations take many forms but when using mapped information it is possible to recognise four methods of approach, which may form successive stages of study. Such methods, or stages, may be undertaken at a variety of levels of understanding and are:

1. The *recognition* or *descriptive* stage. In map studies this can be regarded as the stage at which the model is translated into the map-user's view of reality.
2. The *measurement* stage. Whilst the previous stage is chiefly qualitative in character it is generally followed by the need for precision, i.e. a quantitative approach must be made, in which some of the techniques outlined in chapter 9 will be utilised.
3 The *relationship* stage. At this point the geographer is concerned with the patterns, locations, interaction and possible inter-relationships suggested by the topographic and other distributions shown on the map.
4. The *explanatory* stage. Finally, theories will be put forward in order to account for the patterns noted and described.

For discussion purposes, these four stages will be regarded as distinct but it is often difficult to separate them, since the mental processes involved are all inter-connected.

Recognition

The translation of the map into the details of the earth's surface which it represents is the starting point of any map analysis or, indeed, of any map reading. The map can only be translated adequately, however, if it has been properly and accurately prepared and only then if the limitations of the map, its scale, etc., are understood by the map-user. Map information is conveyed to the user by means of symbols, words and colour or shading. These must either be adequately explained by means of the map legend or else be completely self-explanatory.

Information on maps is conveyed also by the patterns of lines, shading and symbols and this is the point at which *recognition* and *relationships* can become confused. Thus, a town will be recognised by the name lettering, combination of housing and road symbols and possibly additional symbols such as railway lines, industries and the like. Nevertheless the primary activity will be that of recognition and identification of the main features of the town, the shape of the built-up area, the presence or absence of functions such as roads, rail routes, airports, industries, etc. Similarly, other geographical features may be recognised by simple relative patterns, e.g. contour lines and water lines will give information about drainage patterns, river-basins, watersheds, etc. There is thus always an element of interpretation or assessment of relationships in the simplest descriptive recognition of the physical or human landscape (fig. 10.1).

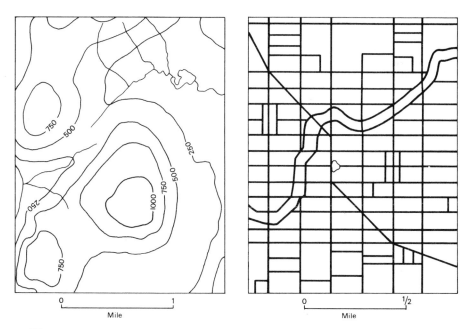

Fig. 10.1. Some simple linear patterns. (a) Contours and drainage network (b) Routeways and settlements. (a) is taken from Ordnance Survey One inch map of part of Skye, (b) is based on a street map of Christchurch, New Zealand.

Measurements

In order to carry out more precise methods of geographical analysis it is necessary to go beyond the initial and qualitative descriptive stage to that of quantification. A number of measurements that can be made from maps have been described in chapter 9 and are summarised in fig. 10.2. Such measurements will be used in conjunction with the description of the mapped topography of stage one so that the map-user can build a detailed picture. Measurements will include details of shapes and possibly also of rank. A distinction may also be made between the static features of the topography and the more dynamic features presented on certain thematic maps. Nevertheless, the vast majority of maps present only information that is correct at a particular moment in time, i.e. they are *synoptic* in character.

Measurements made from maps will need to be recorded in some way: it is most likely that this will be in a tabular manner. The data will

Measurement type	Landform Data	Cultural Data
Single values (by direct reading or interpolation)	Spot heights Absolute height Relative height Stream order	Population figures Production data
Linear values (by 'scaling' against map scale bar)	Stream lengths Coastline lengths Watershed lengths Profiles Water flows Currents	Distances, either direct or route distances between selected features, towns, etc. Movements and flows Vectors
Areal values (by planimeter, geometry or sampling)	Drainage basins Hypsometric data Landform units Rock outcrops, soil associations	Land use data Administrative regions Percentage and density data
Angular values (by direct measurement or calculation)	Slopes Aspect and orientation	Movements – population communications products, trade, etc.

Fig. 10.2. Some selected measurements types in landform and human (or cultural) geography.

then be available for statistical manipulation and further computations. It is also likely that some mapped information will still need to be compared with other records for the same locality, i.e. its most significant factor will be its position relative to other points. The map co-ordinate system of locating points is therefore most important and may be incorporated with other tabulated information. If a considerable amount of information is to be recorded against a host of carefully located points, the most convenient way of assembling this information will probably be with some form of 'digitising' equipment (fig. 10.3). Additionally, a variety of complex information can often be presented in a tabular form as a *matrix* (fig. 10.4).

Relationships

Since the geographer's task is frequently that of describing and analysing the inter-relationships that make up the landscape many of the descriptive and quantifying tasks mentioned above are useful tools for the map interpretation of an area, large or small.

Fig. 10.3. Digitising. Each change in direction of the line is recorded in terms of 'x' and 'y' co-ordinates which can be stored in a data bank and the line subsequently reproduced at any convenient scale. Alternatively, change in direction can be recorded from a curved line by regular interval spacings at which 'x' and 'y' co-ordinates are recorded.

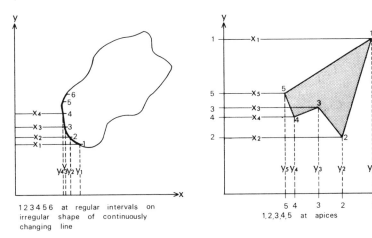

1 2 3 4 5 6 at regular intervals on irregular shape of continuously changing line

1,2,3,4,5 at apices

(a) Relative relief by 500 m squares, part of the Luxembourg Oesling, Trois-vierges area (in metres)

30	30	25	35	25	15	15	60	40
20	25	15	20	15	25	30	55	25
20	15	5	30	40	25	40	50	50

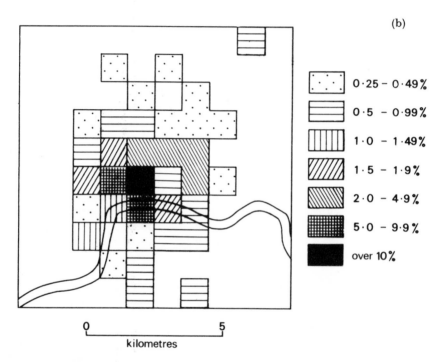

(b)

0·25 – 0·49 %

0·5 – 0·99 %

1·0 – 1·49 %

1·5 – 1·9 %

2·0 – 4·9 %

5·0 – 9·9 %

over 10 %

0   5
kilometres

Fig. 10.4. (a) Section of table of data relating to grid squares, forming a matrix of information based on a map study in the Luxembourg Oesling.
(b) Cartographic representation of data collected on a grid square basis. (Percentage of workers employed in printing. By courtesy of J. E. Martin.)

In very many cases the interpretation of the mapped data is a case of detecting *patterns* exhibited on the map. Such patterns may be in the form of lines (e.g. iso-lines or communication linkages) or it may be from patch or areal data (e.g. land use, buildings) as described in Chapter 3.

Simple patterns of either a linear or areal type are easy to establish and some possible examples are given in fig. 10.1. More complex patterns occur when a variety of information is presented and it is often desirable to simplify the map for purposes of analysis. This may be attempted in one of the following ways:

1. *A 'sieve' method.* Since the map (in particular, the topographical map) is composed of an overlay of several sets of information, usually in different colours, it is relatively easy to disentangle these sets by simple tracing or even by obtaining prints of the map of only the particular colour in which the student is interested, e.g. 'water and contour' prints showing only the brown and blue printing. These are sometimes easily available (Ordnance Survey and some European map series). Similarly, the total built-up area, distribution of particular land uses can all be extracted and shown separately (see the various Land Utilisation memoirs for examples of this method) (fig. 10.5). Possible relationships can thus also be considered if two or more of these extracts are combined.

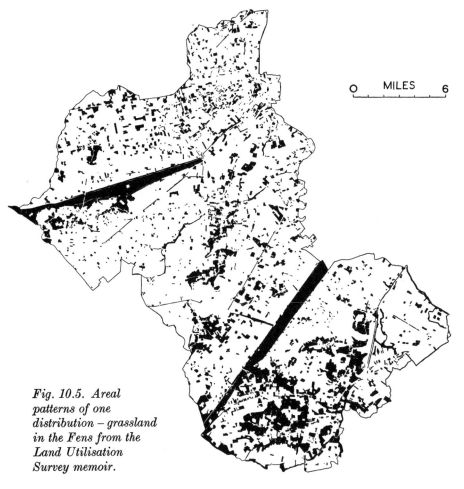

MILES

O   6

Fig. 10.5. Areal patterns of one distribution – grassland in the Fens from the Land Utilisation Survey memoir.

2. *Generalisation methods.* Frequently the detail of a particular distribution, and hence its pattern and relationships with other features, is concealed by the complexity of its occurrence, number of isolated patches, etc. In this case some form of generalisation may help to clarify the picture and probably the most well known example of this type is that which has been used for 'generalised contours'. The technique is applied in order to remove minor erosional landforms so that major forms can be distinguished (fig. 10.6). Such studies can be highly subjective in approach unless great care is taken, e.g. the application of some geometrical rules in order to distinguish 'bevelling' or the drawing of 'generalisation lines' tangential to concavities in the detail lines.

On the other hand, using point data extracted from the map together with the point co-ordinates it is possible to re-create a 'surface' to which these data can refer, either factually as with erosion levels or hypothetically as with economic data, production, population statistics.

An alternative to this is the method of Trend Surface Analysis in which the point data is compared with mathematical surfaces of varying complexities and the degree to which these 'fit' can be seen and the displacements or 'residuals' measured (fig. 10.7).

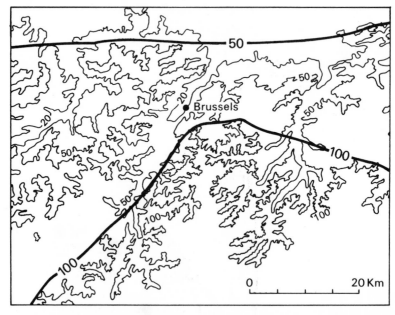

Fig. 10.6. Method of constructing generalised contours. The base map is part of the small scale relief map of Belgium (1/500,000) which shows the complexity of the contour pattern. Regional patterns of relief can be appreciated by generalising the contours as shown.

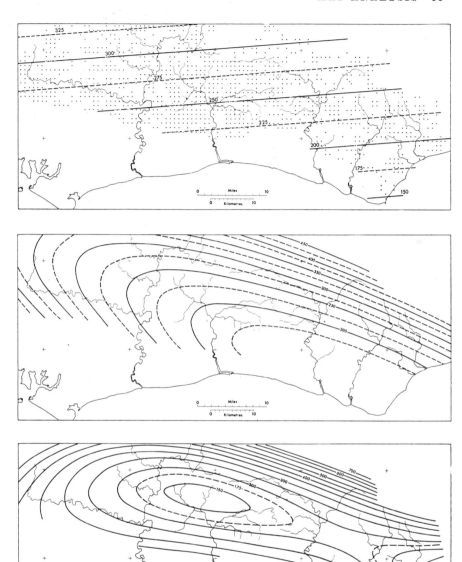

Fig. 10.7. Example of trend-surface maps. Computer calculated data from summit levels in the Sussex Weald compared with maps of theoretical or algebraic surfaces which approximately coincide with reality (after Thornes and Jones 1969). Surfaces shown by isolines representing (top) a linear function, (middle) a quadratic function and (bottom) a cubic function.

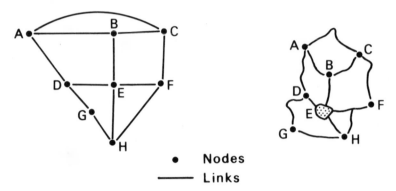

Fig. 10.8. Nodes and links. Left hand pattern is the simplified topological one, whilst that on the right shows the geographical reality – town with adjacent villages and roads linking these settlements. This pattern can be summarised in terms of 12 links, 8 nodes (or vertices). Also, each point can be listed in terms of its number of linkages. Thus A, B, C, D, F and H all have three links, E has four and G only two.

3. *Sampling methods.* In order to assess the occurrence of data, particularly land use information or distributions which can be considered to cover substantial areas either individually or in aggregate on the map it is useful to take a series of samples at various points on the map. Sampling techniques are varied but again normally make use of the grid referencing system on the map together with a pattern of, e.g. random sample points (fig. 9.7). Such a technique has been used with land use data where it is required to find out percentages of each of a group of usages. (Frolov and Maling 1969). Many other methods of studying relationships from mapped data are now available (see Harvey 1969 and Haggett & Chorley 1969) and stem from the realisation that this information can be treated in mathematical ways that are not confined to those of the geometrical distributions of the topography. Thus the techniques of topology, considering nodes and links, are of particular interest (fig. 10.8). The value of different map projections when considering distributions on small scale maps, leads to other possibilities of analysis.

Part IV

Making maps

SYMBOLS
COLOUR CONVENTIONS
LETTERING STYLES
DRAUGHTING PROCEDURES
MAP REPRODUCTION

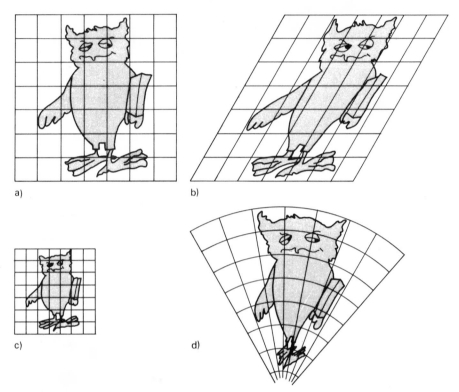

a)

b)

c)

d)

Fig. 11.1. Variations in shape of a given outline as a result of transference to grids of different patterns. (a) and (c) show the use of grids at different scales in order to re-draw outlines and detail at different scales. Method (b) or (d) would be used in re-drawing on a different projection. Method (c) provides an approximate way to re-draw at a different scale where photographic or optical methods are not available. In all cases the draughtsman must transfer line positions from old to new squares by visual estimation of intersections with grid lines, drawing more grid if necessary. In addition, linear distances can be scaled by means of pro-portional dividers.

11 Mapping techniques

The exact methods by which a presentation will be made of geographical information depends on:

1. The type and form of the information.
2. The region or geographical area under consideration.
3. The purpose of the presentation.

The range of possible methods is therefore very extensive, and the purpose of this chapter is to indicate possible routeways towards the cartographic presentation of data.

The raw data for the map will be gathered from a number of sources, ranging from existing maps, statistical tables, written reports, guides, etc., to air photographs and computer print-outs. This data will be capable of being mapped as long as it refers to precise locations which can be either points, lines or areas (p. 30).

As has been seen before (p. 7) a map can be regarded as being either *topographic* or *thematic* in type. Most of the following discussion is concerned with thematic mapping methods and, more particularly, with the preparation of cartograms, sketch maps and maps to be used as illustrations or visual presentations in geographical studies. Technicalities and drawing procedures will be considered in chapter 12.

The first essential in any mapping operation is the preparation of the framework, i.e. the mapping procedure follows one of the golden rules of surveying – 'work from the whole to the part'. The framework may take several forms and may in fact be that of an existing topographic map either to which data is to be added or from which some data is to be extracted, or a combination of these two. In many cases a basic framework may have to be constructed from existing maps and this will involve changes in scale or in projection. Both of these changes may need to be computed in order to draw the grid but the transference of detail will be by some visual method (fig. 11.1). Detail is transferred by inspection from the square or geometrical shape of the original to the corresponding square or shape in the new grid. If necessary, additional shapes or squares will be drawn to subdivide those shown; diagonals and estimations of intersections with grid lines may also be used (Guest 1969, Monkhouse & Wilkinson 1971).

As mentioned above, the cartographer will frequently need to re-draw part of a map at a different scale. This can be done by a visual method or by one of the following:

a. Photographic processes, by direct photography of a map and its printing at an enlarged or reduced scale as required.
b. Episcope methods, in which a tracing is made of the image at the required scale. Special equipment is available (e.g. Grant projector, Plan Variograph) by which the original document is projected on a horizontal glass surface and the image enlarged or reduced as required. This can then be traced or photographed (fig. 11.2).

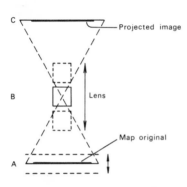

Fig. 11.2. The optical method of changing scale: principle of the Grant projector, Plan-Variograph or Episcope. The drawing to be enlarged or reduced is placed on the carrier plate at the bottom of the diagram and the illuminated image is projected on to a drawing table. This may be by back-projection on to a sheet of tracing paper or film on a glass surface, but in the episcope the image is projected downwards on to a drawing surface. The lens position is adjustable and in the Grant projector so is the position of the carrier plate.

However, if no suitable existing map is available or if the base maps are on different map projections (Steers 1965) to the required map, the cartographer's first task will be the preparation of a framework (possibly by re-drawing on a new grid – see above and fig. 11.1). If the information is made available by tabulated values referring to a rectangular grid this must be carefully drawn up on the drawing board or, if precision to fractions of a millimetre is required, by using equipment such as the *co-ordinatograph*.

The above information will be required both in the initial stages of mapping in order to assist the cartographer in the location of detail and also when the map-user is extracting information from the map. It is, however, background information and should be kept to the minimum so as not to provide too much 'noise' and obscure the main theme of the map. This can be done by using thinner lines or neutral colours, i.e. the locational data must be made of secondary importance. It can consist of one or more of the following:

1. A grid system – usually rectilinear;
2. Outlines of natural features – coastlines, rivers, etc.;
3. Outlines of land-use features – built-up areas, land parcels, etc.;
4. Boundaries of administrative divisions;
5. Significant routeways;

and will also frequently include some place names, grid reference values, spot-heights, etc.

The primary thematic data of the map, i.e. the information message which the map is to portray, is presented in a bolder way than the background, secondary information. This can be achieved in the form of heavier linework, the use of shadings, colour, etc.; and the addition of symbols (fig. 11.3). Explanatory lettering may be printed on the surface of the map but is more likely to form an extensive legend. Included in this category are the various graphs, diagrams, etc., which may be printed on the map and which can be viewed as appendices. These can include tables of data, short sections of text, or graphs of statistics relating to specific places or areas shown on the map. Such insets are referred to as cartograms but may themselves be diagrammatic map representations which is another use of the same term.

It is therefore possible to distinguish the following problems in the making of thematic maps:

1. Considerations of map content. Together with content must be considered the map's purpose and aim.
2. Methods of presentation. Under this heading are the factors of map scale, layout, legend, representation, etc.
3. Map style. The form of linework, lettering types, symbols, colour shades, map shape, etc., make up this section.

When decisions have been made on these three points the detailed map specification has been made.

Map content
This has already been discussed in general terms and it has been shown that two main categories of information are represented on thematic maps – the background information and the message or data that the map is to present. The aim of the map and the audience to which it is directed must be considered also at this stage.

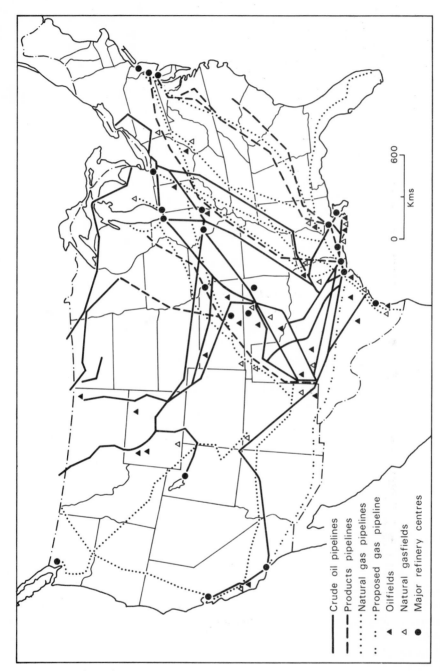

Crude oil pipelines
Products pipelines
Natural gas pipelines
Proposed gas pipeline
Oilfields
Natural gasfields
Major refinery centres

0   600
Kms

Fig. 11.3. Example of a thematic map in which information is portrayed by variations in style of line symbols. Pipelines in the United States (Philips Geographical Digest 1970).

Presentation
Most of the possible methods of data presentation have been considered
in chapters 2 and 3 and are summarised in fig. 11.4.

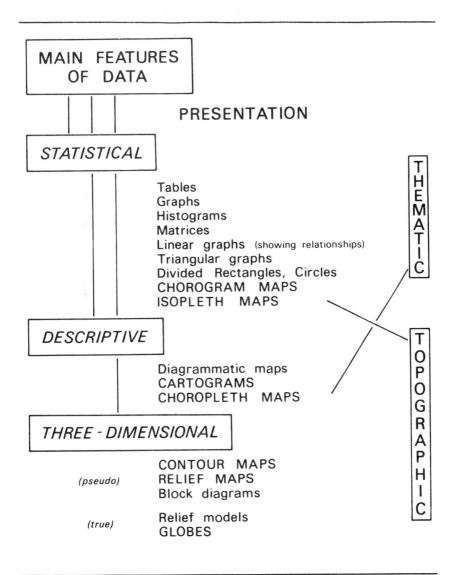

Fig. 11.4. Table of methods of data presentation.

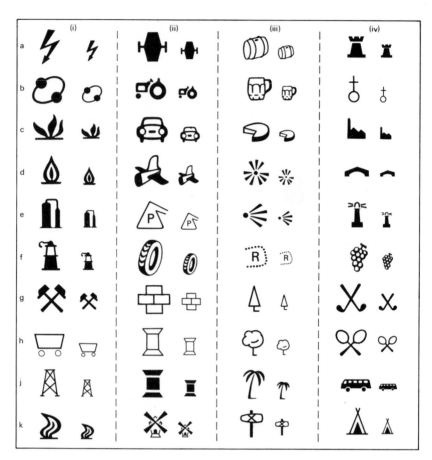

Fig. 11.5. Examples of some of the many symbols available for thematic map use. Each symbol has been shown in two sizes in the ratio 3 : 1; they represent the following features:

Column i
(a) electricity production
(b) atomic energy
(c) thermal electricity
(d) gas production
(e) petrol refinery
(f) coal
(g) iron
(h) non-ferrous metals
(j) petroleum
(k) natural gas

Column ii
(a) iron and steel production
(b) agricultural equipment
(c) motor-car construction
(d) aircraft construction
(e) petro-chemical industry
(f) rubber industry
(g) brickworks
(h) cotton and linen
(j) wool
(k) mills (grain and flour)

Column iii
(a) wine production
(b) brewery
(c) dairy produce
(d) panorama
(e) viewpoint
(f) nature reserve
(g) conifers
(h) leafy trees
(j) palm trees
(k) prehistoric remains

Column iv
(a) fortress
(b) church or cathedral
(c) important factory
(d) bridge or viaduct
(e) lighthouse
(f) vineyards
(g) golf
(h) tennis
(j) bus or coach station
(k) camping site

The graphical form in which the data is to be presented determines the amount of detail that can be shown. In the representation of statistics, geometrical shapes are generally used and the entire figure may well be a geometrical pattern – as in various forms of graph (fig. 9.13) or topological diagram (fig. 10.8). The superimposition of graphical shapes on a map outline leads to other diagrammatic forms (figs. 9.12 and 10.7). Examples of thematic maps are almost limitless (figs. 1.4, 3.2, 4.3, 10.4, 10.5, 10.6, 11.3, 12.4) but, in general, as scale increases a greater amount of detail can be shown.

The representation of detail on diagrams and maps varies with the type of information to be shown as well as with scale. At the largest scale it is usual to show all information in plan form together with explanatory names and/or abbreviations where necessary (e.g. official 1/1,250 or 1/2,500 plans). With decrease in scale outlines become generalised but names and abbreviations will still be shown as far as possible (e.g. 1/25,000) until all information is symbolised (as at the topographic map scales of 1/50,000 or 1/63,360). Thematic maps are sometimes prepared at large scales and will largely conform with these remarks, but the majority of distribution maps in which a shading represents quality and/or quantity of presence are at smaller scales (e.g. 1/250,000 or smaller). Choropleth and isopleth methods characterise such maps although these techniques may be used at all map scales.

Map style

Although the two previous factors will to some extent control cartographic details of style, it is necessary to make decisions on the following points:

1. Symbols
2. Linework widths
3. Colours and/or shading.

Before an exact decision can be made on some of these points the scales of both the final map and that of the original, or fair drawing, must be decided. Normally the map will be drawn at a linear scale of $1\frac{1}{2}$, 2 or even $2\frac{1}{2}$ that of the final map. All symbols, widths and lettering sizes must be chosen with the appropriate reduction factor in mind (fig. 12.4). This drawing will be reduced photographically to the correct reproduction size.

Map symbols (fig. 11.5) will usually conform to conventional usage and can be either qualitative or quantitative. In the latter case, considerable care must be taken in their drawing, as the exact size of the symbol will have a precise meaning. Pre-printed symbols or a stencil

1. BLACK for details of settlement and ground plan-lines, point information, lettering and topographic features such as rock outcrops also, for boundaries.
2. Strong BLUE for drainage network.
3. Reddish BROWN for contours and also possible for rock outcrops. (Grey can also be used.)
4. Strong GREEN for vegetation, tree and woodland symbols.
5. GREY for shading or hatching lines of built-up areas, buildings.
6. YELLOW for any distributional tones required or additional boundary lines (as zone alongside lines).
7. Dark BROWN for additional detail.
8. Strong RED for communications network.
9. Light BLUE for areas of water, seas, lakes.
10. Light RED available for higher relief shading.
11. Light GREY for relief shading.

The first three can be considered as the major colours of the linework on the topographic map.

These colours will show through those of the next group (4, 5 and 6) which serve to complement 1, 2 and 3. The remainder are supplementary colours.

Note that 1, 2, 3, 4, 7, 8 are strong colours, to be used for lines and symbols in the main.
5, 6, 9, 10 and 11 are light colours, to be used chiefly for shadings but possible for linework.
6, 9 and 10 can be used together and in various strengths to produce a range of shadings.

Blue and blue mixtures are generally regarded as *cold* colours – use these therefore in arctic and alpine areas, e.g. icefields, higher relief.

Red and red mixtures are generally regarded as *warm* colours.

Fig. 11.6. Table of colour usage for different phenomena on maps.

sometimes can be used to maintain uniformity, but it is frequently necessary to construct symbols geometrically (Monkhouse & Wilkinson 1971, Hodgkiss 1970).

Widths of linework will vary on the map as between lines of background information and those of greater importance; grid lines are usually kept to a minimum and use will be made of various line styles (figs. 11.3 and 10.1), especially if the map is in a single colour.

The use of colour leads to specific problems, notably in the preparation of the map originals. Colour printing of maps is usually carried out by a 'separation' process in which a separate printing plate is prepared for each colour. These separate drawings must be made on stable material so that they exactly agree in size and shape. Also, 'register marks' must be drawn on each separate drawing to assist in this positioning over one another (fig. 12.4). When colour tones or shadings are involved, care must be taken to ensure that appropriate gaps are left in other colours so that 'overprinting' does not occur; this may be desired, however, in some cases, depending on the colours involved.

The choice of colours is always a vexed one. Considerations of conventional usage, the map type, and generalities of appearance and aesthetics are all involved (Keates 1962). However, a number of widely accepted conventions can be recognised and are summarised on fig. 11.6. In addition, printing problems will make it necessary for any cartography involving colour maps to be a co-operative matter between map designer, cartographer and printer.

For uniformity of shading on any map it may be necessary to use pre-printed tones or symbols or stencils (chapter 12), often causing some limitation in the range of shadings available to the map designer. If a series of maps are under consideration then additional care is needed in the choice of both colours and shadings, to preserve uniformity and to allow for the greatest possible range in data.

The numerous lettering styles and type faces available today provide the map-designer with considerable choice (fig. 11.7). Uniformity in lettering may be achieved by using stencils (fig. 12.7) or by applying pre-printed letters and numbers. This is desirable in maps which form part of a series (i.e. one sheet of a number covering adjoining regions) or a sequence (i.e. several maps showing different facets of the same area) as well as those which are to be reproduced in one volume. Care must be taken to allow for any reduction factor between fair-drawing and printing. If a different reduction factor is used this will change the values of shading as well as of heights of lettering. Clarity and uniformity will make for easier reference and reduce the 'noise' factor. Usage on some topographic maps has led to the following conventions, which are *not*, however, universal.

Upright lettering for town and place names.
Italic lettering for natural features, seas, rivers, etc. (or for areas and regions as opposed to settlements).
Gothic or Old English lettering for landscape features of antiquity.
Ordnance survey usage is indicated in fig. 3.6 and Appendix I.

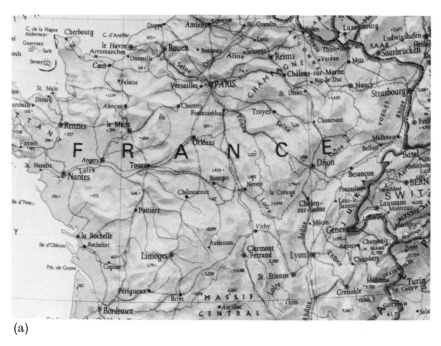

(a)

(c)

(b)

Fig. 11.7. Some different lettering styles and the resulting changes in general appearance of part of an atlas map.

(a) *Mixture of* Albertus *and* Joanna *italic type. Country names are set in Univers Medium. The Joanna italic has narrow serifs and no variations in 'weight' of lettering are used on this map. Differences are emphasised by changes of point size or of typographical style. Thus towns in different Joanna italic, regions in Albertus.*

(b) *Type on this map is Univers throughout. Country names are in Extra Bold (stippled) and a wide range of weight in the Univers type is used, from the very light river names to the extra bold of the towns.*

(c) *All lettering is in Univers type. Four weights of roman are used for town names and italic for physical features, rivers and spot heights.*

12 Draughting techniques

In this chapter a brief consideration will be made of basic methods for the preparation of maps, diagrams and similar illustrative material for either publication or fair presentation. The cartographer must therefore be a draughtsman and although there are many aids to drawing a minimum level of skill must be obtained. This can only be reached by practising individual methods and styles since map preparation involves the cartographer in acquiring an expertise in drawing, lettering and the layout and presentation of material. It is necessary therefore to consider both the 'tools of the trade' and their application to cartography.

Drawing surfaces
For best results it is necessary to use good materials. A considerable range of drawing materials are available from which the cartographic draughtsman will choose according to the type of job in hand. These materials include the following:

1. Paper and board (opaque);
2. Tracing paper and tracing linen (translucent);
3. Plastic film (opaque, translucent or transparent);
4. Enamelled surfaces, usually on metal, e.g. zinc (opaque).

Drawing paper is the least stable of these materials because it is liable to expand or contract with atmospheric changes, especially those of moisture. Tracing linen and tracing paper display the same problems but thin board, e.g. Bristol board, is more stable because of its thickness. It cannot be rolled or bent, however. Paper, board and cloth can be defined by their thickness characteristics, generally quoted in decimals of an inch or millimetre, but it is common to encounter descriptions in terms of:

(a) Paper type and quality, e.g. cartridge paper, Engineers, Kent made.
(b) Drawing surface, e.g. smooth, glossy, art or hot pressed, rough.
(c) Weight, e.g. 80 lbs – referring to the final press weight in the manufacturing process.
(d) Sheet size by terms such as Imperial, Demy, but these are currently being replaced by International sheet sizes (A0, A1, A2, etc.). Details are given in Appendix III to the Royal Society Glossary (1966).

In the case of paper, these descriptive terms are usually combined in the total nomenclature, e.g.

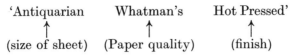

Plastic film comprises a range of drawing materials of different characteristics:

1. Clear film – transparent material.
2. Opaque film, the surface of which can be cut or 'scribed' with clear lines.
3. Opaque film, covered by surface capable of being 'peeled' to give clear areas.
4. Translucent material.

A perfectly clear film is required for the preparation of display material for projection purposes, e.g. overhead projector transparencies. Translucent or tracing material is that most commonly used as a general purpose plastic film in the preparation of maps when stability is important and is marketed by several commercial organisations under such trade names as 'Kodatrace', 'Ethulon', 'Permatrace'.

Surfaces which can be marked with lines by a fine steel or similar pen (fig. 12.1) ('scribed') provide stability and also allow of greater uniformity of linework than that given by applying ink. A commercial production of this is 'Scribecoat' and is usually red or orange in colour but may be specially prepared in other colours. Similar colours are used in the production of 'Peelcoat'. In this material an upper surface of plastic can be removed by first cutting round the desired shape and then removing by peeling (fig. 12.1). In this way areas to be shaded can be indicated in a negative manner or other parts of the map cleared to leave an opaque area where the shading or colouring is to appear (positive).

Drawing equipment

Most cartographic work is undertaken by drawing black or coloured outlines on a white background and drawing equipment is therefore needed to apply these lines or colourings. For sketching purposes, preliminary layout work, etc., pencil lines will be used whilst final linework will be in ink. Fair drawing on scribing material is carried out with a special scribing pen which uses a sapphire tipped or steel point. Various designs of nib point allow for variations in width and style of line in both ink work and scribing.

Most line work will be aided by the use of guide-lines, templates, stencils, etc., to ensure accurate geometry. Thus, straight lines are

E

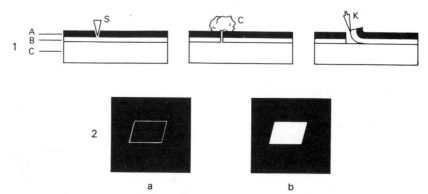

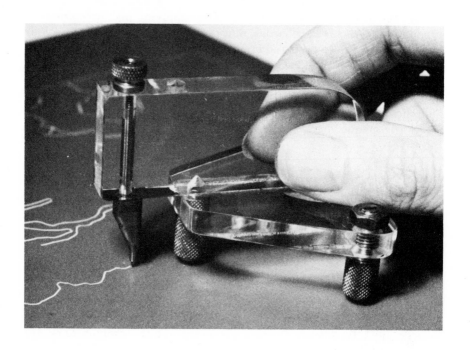

Fig. 12.1. Upper diagram (1) illustrates the use of Peelcoat drawing material. A is the opaque material, B a soluble adhesive and C the transparent film. S indicates the scribing point, C the dampened sponge and K a knife point used to remove the Peelcoat; diagram (2) shows a simple outline before, (a), and after, (b), removal of opaque Peelcoat. The lower picture shows a scribing tool in use.

drawn against a machined straight edge, parallel lines drawn with a parallel ruler, perpendiculars and angles drawn by reference to a protractor or set-square. Many of these can be assisted by placing a sheet of accurately squared paper (graph-paper) beneath the tracing paper (fig. 12.10) and/or by the use of a drawing table fitted with parallel guides, protractor, etc. (fig. 12.2).

Ink line work is the most commonly used method in cartographic draughtsmanship and the following pens and inks will probably be encountered:

1. Stencil pens, e.g. UNO, Leroy, Standardgraph, Rotring Varioscript.
2. Line pens, e.g. Ruling pens, Rapidograph, Rotring Variant.
3. Freehand or quill pens, including special lettering pens and nibs.
   Some examples of pens, compasses and their use form fig. 10.3.

All pens must be *filled* with ink and not dipped into an ink bottle. In this way ink will only be placed in the reservoir and not on the outside of the nib. It is necessary to ensure both a steady flow of ink and to reduce contact friction. The pen should be slightly tilted in the direction of movement. Pens attached to compasses are similarly tilted and in all cases movements should be made in a steady and deliberate manner.

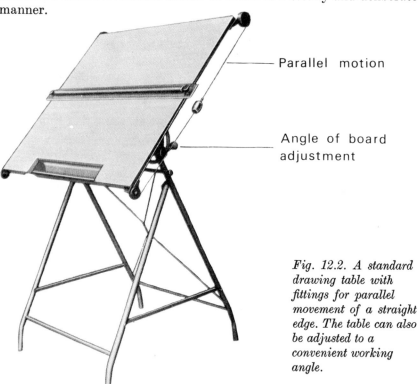

Parallel motion

Angle of board adjustment

Fig. 12.2. A standard drawing table with fittings for parallel movement of a straight edge. The table can also be adjusted to a convenient working angle.

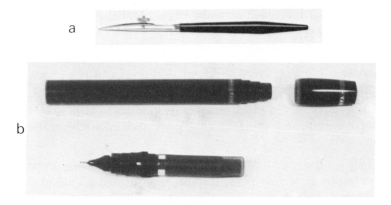

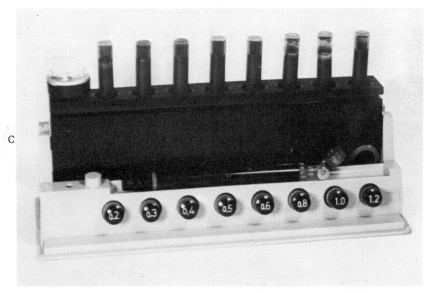

Fig. 12.3. Pens for ink linework, lettering with stencils and ink compasses.
(a) Ruling Pen.
(b) Variant drawing instrument (Rotring).
*(c) Rapidomat stand for Rotring pens (Varioscript and Variant types). This
is a humidifier which ensures that pens are kept ready for use; the pen caps are
arranged in front with the line width dimensions (in tenths of a millimetre)
visible. Pen holder and attachment for stencil use are also to be seen.*
(d) Line thicknesses drawn by the various nibs in the Rotring range.
*(e) Detail of the difference between the nib points. Variant point has a 'shoulder'
to prevent ink from flowing under guide edges against which it may be used.
Varioscript has no such shoulder as it is used vertically in stencils for lettering.*
(f) Using the Variant as a freehand drawing instrument for uniform linework.
(g) A drop compass for drawing very small circles.
(h) Ink compass in use showing reservoir and fitting.

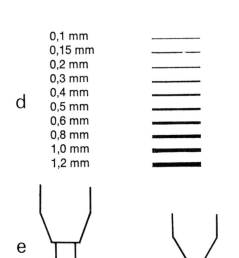

0,1 mm	
0,15 mm	
0,2 mm	
0,3 mm	
0,4 mm	
0,5 mm	
0,6 mm	
0,8 mm	
1,0 mm	
1,2 mm	

d

e

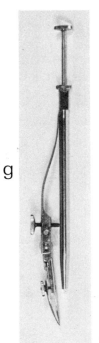

g

f

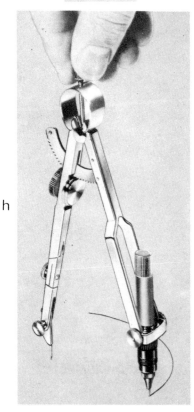

h

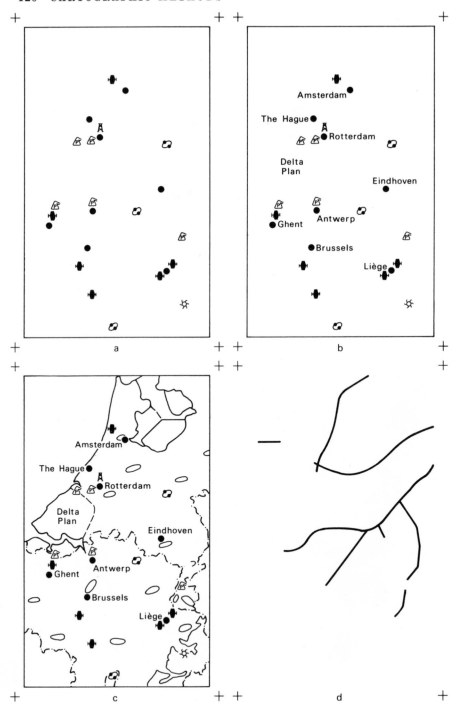

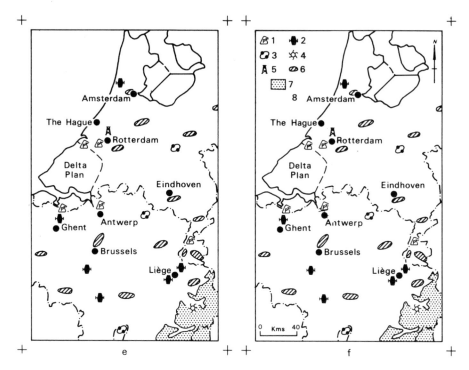

Fig. 12.4. Stages in draughting a map. The diagram on this and the facing page show a suggested series in which a two-colour map might be drawn.

(a) *Overall map size determined and border drawn (probably in pencil only, for inking over at stage (f). Symbols positioned. Register cross marks drawn outside map frame at all four corners.*

(b) *Lettering added.*

(c) *Linework (coast, frontier and edges of areas to be shaded) added.*

(d) *Separate drawing made for second colour (pipelines) with identical register marks in corners. Gap made where line crosses other detail.*

(e) *Shading added to first colour drawing. Note that, when using pre-printed dry transfer lettering together with pre-printed shading, care must be taken to prevent the latter adhering to the letters and removing them. It may be prudent in some cases to apply the shading before the lettering and cut 'windows' where the lettering will be placed. If line shading and stipple are hand drawn, this precaution will not apply.*

(f) *Add legend, scale, north point and title if required.*

MAP LEGEND:

1. *Petro-chemical industry*
3. *Nuclear power electricity*
   generation
5. *South Holland oilfield*
7. *Land over 400 m*

2. *Iron and steel industry*
4. *Pumped storage electricity*
   generation
6. *Main industrial areas*
8. *Natural gas pipeline (in second*
   colour)

Procedure (fig. 12.4)

A systematic approach is necessary in the drawing of a map and, although a possible series of stages is outlined below, the individual cartographer will proceed in an orderly way most suited to the types of work he is required to undertake.

1. Designing the map. In many cases the draft map will be prepared from some existing map and will probably be drawn in a variety of inks or pencil on tracing paper. The final 'fair drawing' is prepared by tracing over this on transparent or translucent material. In some cases the draft may be made in very light pencil on a good quality white surface, e.g. Bristol board, and this may be inked over with great care. The precise size of the draft map will depend on that of the final reproduction. These can be correctly related by simple geometry, as shown in fig. 12.5.
2. Add single symbols. These may be qualitative, e.g. tourist sites, or quantitative in which the size of symbol gives an indication of town size, industrial production or other activity. Lettering and linework must therefore conform to these symbols and thus the symbols will be drawn first.
3. Applying lettering (see below).

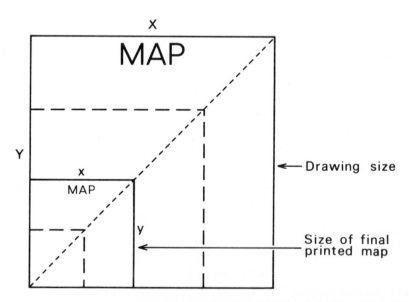

Fig. 12.5. The relationship between the final reproduction size of the map and the size at which it is drawn are easily obtained by extending a diagonal as shown here. Also note the effect of reduction on lettering size. Other possible sizes of original or final print are shown by the broken lines.

4. Line-in all solid and broken lines, making gaps where lettering has been placed and allowing for any break due to shading or symbolism still to be added. Work from the finest line to the heaviest if there is doubt as to their distinction in the final map – lines can easily be thickened but the removal of ink is tedious. Complex linear symbols may be applied using a pre-printed adhesive tape produced under trade names such as *Chartpak, Normatape, Letratape* (fig. 3.1). Erasures can be made by various methods:

(a) India-rubber/abrasive erasers – hand-held or electrically operated.
(b) Abrasive paste.
(c) Sharp cutting edge, of knife or razor blade.

It is usually necessary to restore the drawing surface of the paper, film, etc. This can be done by burnishing the area of the erasure with a smooth tool of metal, hard plastic or ivory.
5. Apply tones, shadings, colour washes, etc. (see below).
6. Add margins, titles, legend, decorations, etc., not included in basic lettering at stage 3.

Lettering
Lettering can be added to maps and diagrams either by careful freehand work or by using one of several aids to lettering, such as stencils or pre-printed letters.

Freehand lettering may consist of single strokes or be 'built-up'. Guidelines are essential to control both the height and the character of the letters, e.g. angle of slope. Guidelines are often desirable in applying lettering by stencil or other methods and should always be used to control blocks of text, legends, etc. A sheet of squared paper beneath the draft map being copied is of considerable assistance in lettering as well as shading but must be made 'square' with the map margins. However, the labelling of some features, such as rivers, hills, flow lines, routeways, may need a curving word form and guidelines are then drawn (fig. 12.6).

Stencils of a variety of character heights and styles are available and the types most commonly encountered are:

1. UNO, Standardgraph, etc., where a plastic guide is held above the paper surface and controls the movement of a vertically-held special pen (fig. 12.7). The pen can be obtained in various nib thicknesses depending on the stencil in use.
2. LeRoy system, where a metal or plastic guide controls the movement of a specially equipped pen. The guide in this case is placed parallel to and below the lettering.

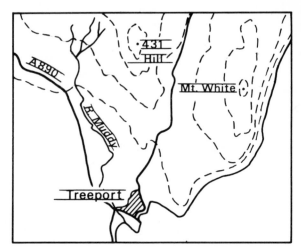

Fig. 12.6. Guide lines for positioning lettering. Note that the writing follows the general line of roads and rivers rather than conforms to all the twists. Also note the position of heights, summit names, and town names relative to the features.

Pre-printed lettering has been used for many years in commercial mapping offices and has now become widely used in drawing offices and small organisations. A variety of lettering styles and sizes are available as well as shades, symbols and colours. Two forms are generally available, both of which are wax-adhexive:

(a) A 'cut-out' variety, in which the individual letters or words are on thin transparent film which is made to adhere to the drawing by pressure. Examples are *Formatt* lettering sheets and the pre-set words produced by photo-lettering machines such as the *Photonymograph*.

(b) A 'transfer' variety in which the individual symbols are released by pressure from a backing sheet and are then applied directly to the drawing. Examples are the founts of lettering supplied in sheets under trade names such as *Letraset, Letter-Press*.

With the first category care must be taken in positioning the symbols after removal of the backing material (fig. 12.9), whilst with the second it is necessary to be careful in the alignment of individual letters released from the fount sheet (fig. 12.10). With this type of pre-set symbol there is also a risk of mishap due to faulty adherence but errors can be speedily rectified. Alignment is aided when working on transparent material by having a sheet of squared or lined paper beneath.

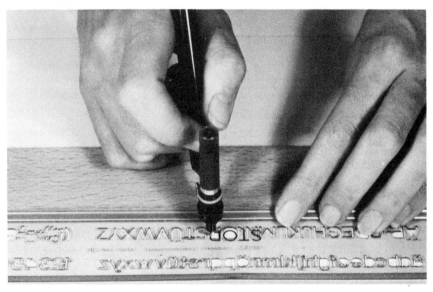

Fig. 12.7. Using a stencil as an aid in lettering. The stencil is a transparent plastic guide held above the drawing surface by its raised edges. In order that the lettering is aligned correctly the stencil must be placed against a straight edge which is held firmly. The stencil pen, in this case a Rotring Varioscript, *is held vertically by means of a special holder or adaptor. The stencil here being used is a* Standardgraph *which contains upper and lower case letters, numbers and some symbols.*

(i)

(ii)

a

ABCDEFGHIJKLMNOPQRST UVWXYZÆŒÇØ
abcdefghijklmnopqrstuvwxyzææç øß1234567890?!(%)&«»;:/+„ˇ^~···°

ABCDEFGHIJKLMNOPQRSTU VWXYZÆŒÇØ 1234567890 abcdefghijklmnopqrstuvwxyzæ œçøß?! (%)&«»;:/+„ˇˇ~···°

b

ABCDEFGHIJKLMNOPQR STUVWXYZÆŒÇØ abcdefg hijklmnopqrstuvwxyzæœçøß 1234567890?!(%)&«»;:/+ˇˇ~°

ABCDEFGHIJKLMNOPQRST UVWXYZÆŒÇØ
abcdefghijklmnopqrstuvwxyzæœ çøß1234567890?!(%)&;:ˇˇ~°

c

ABCDEFGHIJKLMNO PQRSTUVWXYZÆŒ ÇØ abcdefghijklmnopq rstuvwxyzæœçøß ˇˇˇ~ 1234567890?!(%)&«»;:/+

𝕬𝕭𝕮𝕯𝕰𝕱𝕲𝕳𝕴𝕵𝕶𝕷𝕸𝕹𝕺𝕻𝕼𝕽𝕾 𝕿𝖀𝖁𝖂𝖃𝖄𝖅& 1234567890?!& abcdefghijklmnopqrstuvwxyz

d

ABCDEFGHIJKLMNOPQRS TUVWXYZÆŒÇØabcdefg hijklmnopqrstuvwxyzæœçø ß1234567890?!(%)[];:ˇˇ~··°

ABCDEFGHIJKLMNOPQRS TUVWXYZÆŒÇØ ();:ˇˇ~··°~ abcdefghijklmnopqrstuvwxy zæœçø1234567890?!()

Fig. 12.8. (a) *Lettering typefaces, some of the many different alphabets available in a pre-printed form.*

Column i
(a) *Baskerville*
(b) *Times Bold*
(c) *Clarendon Bold*
(d) *Univers*

Column ii
(a) *Futura*
(b) *Times Bold Italic*
(c) *Engravers Old English* ·
(d) *Univers Italic*

ABCDEFGHIJKLMNO
abcdefghijklmnopq

ABCDEFGHIJKLMNOPQRST
abcdefhiklmnorstuvwxyz123

ABCDEFGHIJKLMNOPQR
abcdefhiklmnorstuvwxyz

ABCDEFGHIJKLMNOP
abcdefhiklmnopqrstuv

ABCDEFGHI
abcdefghijk

ABCDEFGHIJK
abcdefhijklmno

ABCDEFGHI
abcdefhijklm

ABCDE
abcdefg

ABCD
abcde

ABC
abcd

AB
abc

(b) *A range of print sizes is shown in the Univers typeface. Note that the size is not that of the actual letter but of the print body upon which it is placed. Left, top to bottom, 8, 10, 12, 14, 16, 20, 24 and 36 point. Right, 48, 60 and 72 point.*

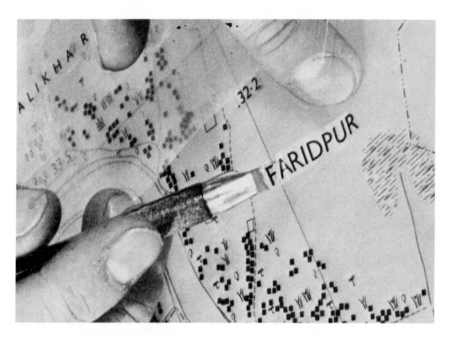

Fig. 12.9. Positioning lettering of the pre-printed-on-film variety.

Shadings, areal symbols, etc.

Both freehand and mechanical methods can be used in applying shadings to maps. Freehand symbols are only possible if exact repetition can be obtained, as for example with lines or geometrical shapes. A useful aid to such freehand work on tracing materials, plastic films and similar transparent surfaces is a sheet of squared (graph) paper placed beneath the map draft. This will provide guidance for parallel lines, perpendiculars, spacing of dots and lines (fig. 12.11).

Symbols are frequently drawn with the aid of a stencil or, where large and geometrically simple, they may be constructed with compass and ruler. Alternatively, use can be made of pre-printed forms, again of both cut-out and/or dry transfer types (e.g. *Letratone, Formatt, Normatone*). Where separate symbols are used in combination to produce a complex shading, e.g. for vegetation, some guide lines will be required – either as an underlay sheet or in light pencil.

Colours

In the preparation of a map which is not to be reproduced, a great variety of colours can be used and the range will only be restricted by the particular needs of the map, legibility and the skill of the cartographer. Water colour paints are most likely to be used in the form of colour washes over larger areas, colour inks over smaller areas.

LETTER-PRESS

```
_ A A A A A A A A A A A A A A      _   a a a a a a a a a a a a a a a a a a a
_ A A A A A A A A A A A A A A          a a a a a a a a a a a a a a a a a a a
_ A A A A A A A A A A A A A A          a a a a a a a a a a a a a a a a a a a
_ B B B B B B B B B B B B B B B B      b b b b b b b æ æ æ æ æ æ æ æ æ   _
_ C C C C C C C C C C C C C C C        c c c c c c c c c c c c c c c c c c c c
_ D D D D D D D D D D D D D D D    _   d d d d d d d d d d d d d d d d d d
_ E E E E E E E E E E E E E E E E E     e e e e e e e e e e e e e e e e e e
_ E E E E E E E E E E E E E E E E E E   e e e e e e e e e e e e e e e e e e
_ E E E E E E E E E E E E E E E E E E   e e e e e e e e e e e e e e e e e e
_ E E E E E E E E E E E E E E E E E E   f f f f f f f f f f f f f f  e e e e e e e
_ E E E E E E E E E  Æ Æ Æ Æ Æ Æ       g g g g g g g g g g g g  œ œ œ œ œ
_ F F F F F F F F F F F F F F F F F F   h h h h h h h h h h h h h h h h ä ä ä
_ G G G G G G G G G G G G G G G G   _  i i i i i i i i i i i i i i i i i i i i i i i i i i i i i i i i i i
_ H H H H H H H H H H H H H H H H      j j j j j j j j j j j i i i i i i i i i i i i i i i i i i i i i
_ I I I I I I I I I I I I I I I I I I I I I I I I I I I  k k k k k k k k k k k k k k k k k k k
_ J J J J J J J J J  I I I I I I I I I I I I I I I I I   I I I I I I I I I I I I I I I I I I I I I I I I I I I I I I I I I I
_ K K K K K K K K K K K K K K K K      m m m m m m m m m m m m m m m m   _
_ L L L L L L L L L L L L L L L L L L L  n n n n n n n n n n n n n n n n n n n n   _
_ M M M M M M M M M M M M M M         n n n n n n n n n n n n n o o o o o o
_ M M M M M   N N N N N N N N N        p p p p p p p p p p p p p p p p p p p p p p
_ N N N N N N N N N N N N N N N N      o o o o o o o o o o o o o o o o o o o o o o o o
_ O O O O O O O O O O O O O O         o o o o o o o o o o o o o o o o o o o o o o o o
_ O O O O O O O O O O O O O O         q q q q q q q q q q  ç ç ç ç ç ç ç ç ç
_ O O O O O O  Œ Œ Œ Œ Œ              r r r r r r r r r r r r r r r r r r r r r r r r r
_ P P P P P P P P P P P P P P P P      s s s s s s s s s s s s s s s s s s s s s
_ Q Q Q Q Q Q Q Q  R R R R R R R      s s s s s s s s s s s s s s s s s s s s s
_ R R R R R R R R R R R R R R R R R R   t t t t t t t t t t t t t t t t t t t t t t t t t
_ S S S S S S S S S S S S S S S        u u u u u u u u u u u u u u u u u u u   _
_ S S S S S S S S S  T T T T T T T     u u u u u u u u u u u u u u u u u u u   _
_ T T T T T T T T T T T T T T T T      v v v v v v v v v v v v v v v ø ø ø ø ø ø
_ U U U U U U U U U U U U U U U U      w w w w w w w w w w w w w w w w
_ U U U U U U U U U  Ç Ç Ç Ç Ç        x x x x x x x x x x x x y y y y y y y y y y y
_ V V V V V V V V V V V V V V V        z z z z z z z z z z z z z z z z z z z z z z z
_ W W W W W W W W W W W W W W         ] ] [ [ ) ) ) ( ( ( ? ? ? ! ! ! Ê É Ê É É Ñ Ä Å Á
_ X X X X X X X X X X  Ø Ø Ø Ø        : : : : - - - . , , , , : : : : : : : : : - - - . . & &
_ Y Y Y Y Y Y Y Y Y Y Y  Ä Ä Ä Ä     § § † † ˚ « « » » ] ] [ [ ) ) ) ( ( ( ? ? ? ! ! !
_ Z Z Z Z Z Z Z Z Z Z Z  Ü Ö Ö       ß ß ck ch á â à ó ò ô ñ í î ì è é ë ê è
_ 1 1 1 2 2 3 3 4 4 5 5 6 6 7 7 8 8 9 9 0 0   ü ö ä ä ä ä ä » » ′ ′ ˝ , , , , ú û ù ø ø ø õ ó
_ 1 1 1 2 2 3 3 4 4 5 5 6 6 7 7 8 8 9 9 0 0   1 1 1 2 2 3 3 4 4 5 5 6 6 7 7 8 8 9 9 0 0
_ 1 1 1 2 2 3 3 4 4 5 5 6 6 7 7 8 8 9 9 0 0   1 1 1 2 2 3 3 4 4 5 5 6 6 7 7 8 8 9 9 0 0
```

5 5 16   5 5 16

Fig. 12.10. Example of a 'fount' or sheet of pre-printed dry transfer lettering. Note that both upper and lower case letters are available, also numbers and some symbols.

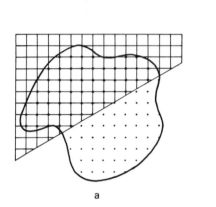
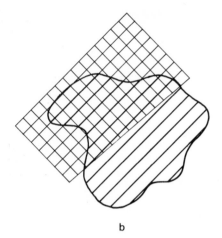

a   b

Fig. 12.11. Using squared paper for guidelines. The squares can be used to aid in the location of lettering, the spacing of a dot shading pattern and also for spacing a line shading pattern.

Preparation for the reproduction of a map in colour will involve the cartographer in working from a monochrome base map showing *all* the information and the drawing of separate 'maps' or 'plates' in black for each colour to be used. This base plate is referred to as the 'key' plate.

It is essential that all these separate plates fit together exactly and for this purpose 'registration marks' are drawn on each sheet (fig. 12.4). These will be carefully lined up at various stages in map reproduction and printing plate preparation: therefore the cartographer must work with equal care in drawing the individual colour outlines. Each outline must be prepared on similar drawing materials, and this must be stable in dimensions. In some cases it may be easier to prepare a 'negative' plate, i.e. filling in those areas NOT to appear in the colour under consideration. An appropriate 'positive' will then be prepared photographically.

13 Map reproduction

The topic of *reprography* is of special importance to the cartographer because it is very likely that he will be preparing maps for publication or reproduction in some form. Also, when handling published maps, the map-user will find that a knowledge of the preparation stages will indicate limitations in the mapped data, for example, accurate measurements can only be taken from precise surveys printed on stable material.

Although the single map in its original form may sometimes be all that is required, multiple copies are often needed. The exact number will depend on the type of map and its degree of sophistication. One can distinguish, for example, between the clearly general interest type of map represented by the tourist topographic map sheet (Ordnance Survey one inch map of Dartmoor, fig. 2.1c) and the specialist diagrams illustrating a paper, thesis or report (figs. 1.3, 3.2, 9.9, 10.7). The former will be published in thousands of copies and can be regarded as 'long-run' printing jobs whilst the latter may require no more than a dozen copies to be made. Between these two extremes are the multitude of topographic and thematic maps and plans used both as separate publications and/or as illustrations in books, journals, etc. The following categories can perhaps summarise the situation:

1. Maps produced in large quantities in which the techniques of conventional printing (single or multi-colour) are economically practicable, e.g. small and medium scale topographic and thematic maps, atlases.

2. Limited publication runs of either separate map sheets or of illustrations in the text of books, journals, etc. These are generally single colour (monochrome) but can be more complex. Normal printing processes are again desirable.

3. Small numbers of copies of individual maps and diagrams.

In the first two cases there is usually a stage of photographic reduction between the original fair-drawn map and the reproduction whereas in the third category 'true to scale' copies are the norm. Similarly, the vast majority of maps produced in small quantities are in one colour, frequently black on white.

A great variety of processes are available for the reproduction of maps and diagrams (Clare 1964 and Hutchings 1970), including the following:

1. Letterpress or 'relief' methods of raised lines to which ink adheres and from which a print is made.
2. Etched or *intaglio* processes of engraving where the ink is carried in a channel. Recessed patterns also form the basis of the *photogravure* process.
3. Level printing surfaces of *lithography* where an inked image is transferred to the paper, generally by means of an intermediate 'blanket' roller.
4. Photographic processes involving chemical reactions to light to produce lines, including 'dye-line' and 'true to scale' copying methods.
5. Reproductions by stencil methods, either waxed stencils and ink or spirit duplicating. The stencils may be prepared by a photographic and/or electronic process.
6. Electrostatic copying methods of Xerography.

The first three techniques are used for 'long runs' of printing, i.e. when a large number of copies is required, whilst the remainder are more likely to be used for smaller numbers, ranging from single copies by electrostatic or photographic methods to hundreds by the stencilling techniques. Where a 'wet' process is involved, e.g. development of photographic film or paper, distortions may occur and accuracy of measurement may be lost. Similarly, lines can become blurred in duplicating and electrostatic copying.

Considering some of these various processes in slightly more detail.

Letterpress printing

Raised symbols, letters and designs have been used for printing since the days of woodcuts. The raised lines can either be placed on a level surface as in 'flat-bed' printing machines, or transformed into a curved printing surface which can be fastened to the rotary cylinder of a high-speed printing press.

The map to be printed must therefore be converted into a printing plate or line block and this is carried out by making a negative of the original reversed as in a mirror image which is then exposed to the printing plate so that the resulting prints will be in the correct sense (fig. 13.1). Where shades or tones are to be reproduced a 'half-tone block' must be prepared. This is made by photographing the original through a 'screen' consisting of a grid of finely spaced lines. Tones are then broken into dots of varying size, depending on the intensity of the tone (fig. 13.2).

Fig. 13.1. *Printing cylinder and impression produced from the mirror image of the printing plate.*

Fig. 13.2. *Magnified example of part of a 'half-tone' printing plate. The plate is prepared by photographing through a screen of fine lines which result in a pattern of squares. However, these tend to produce rounded outlines so in detail the pattern consists of a mass of dots of various sizes.*

Lithographic processes

The original discovery of the basis of this technique of printing was in the late eighteenth century when it was found that an impression could be made from a greased outline drawn on a piece of limestone (*lithos* = stone). Present-day developments of this use the fact that two adjoining surfaces can be printing and non-printing areas if one is made capable of retaining ink and the other receptive to water, greasy printing ink and water being mutually exclusive. The process thus initially depends also on dampening of the paper or wetting of the plate.

The lithographic printing plate is usually of aluminium which is chemically sensitized so that a contact print can be made upon it from the negative of the original map. The print lines are prepared with ink as stated in the next paragraph, and an impression made from these is then transferred to the paper, either directly or, more usually, by way of an intermediate 'blanket' roller (fig. 13.3). The actual printing process is thus not from the actual plate but is 'offset'; the inked plate will therefore be a positive.

The actual preparation of the printing plates involves their coating with a light-sensitive layer which will adhere to the image or printing

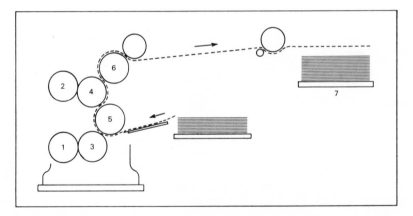

Fig. 13.3. The main features of the offset lithographic printing press. The schematic diagram shows a two-colour press and the direction of movement of paper through the press is arrowed.

1. *Cylinder to which is attached printing plate for first colour.*
2. *Ditto for second colour.*
3. *Blanket cylinder transferring first colour image to 'impression cylinder'.*
4. *Blanket cylinder for second colour.*
5. *Impression cylinder.*
6. *Transfer cylinder.*
7. *Pile of printed sheets.*

areas after exposure to the original (on clear film or similar material). The non-printing areas will then be dissolved away. Plates are coated by a centrifugal method; the plate is placed in a 'whirler', a solution is poured over its face and an even covering is achieved by spinning the plate at a high speed together with controlled air and temperature conditions for drying.

Engraving

Recessed or engraved lines on a printing plate retain ink whilst the smooth surface is wiped clean; the image of the lines, lettering, etc., is then transferred to the paper. This method was used in many early maps, the engraving being carried out by hand on copper plates (fig. 2.10). Early Ordnance Survey maps were so prepared and also Hydrographic Department charts until comparatively recent times. The advantage as far as charts were concerned lay in the fact that amendments could be quickly carried out; the engraved line was relatively easily removed and new information added without destroying the entire sheet. Ink is also retained in depressions on the surface of the plate in photogravure methods of printing shades. The deeper the shade to be printed the deeper must the recessed cell be, so that a greater amount of ink can be held. These cells form a regular pattern over the printing cylinder and after this has been inked a scraper blade removes the surplus from the non-printing surface.

Dyeline reproductions—(Diazo processes)

When a quick copy of a single colour line map is required and provided that the original is on a transparent or translucent material such as tracing paper or film, it is possible to make a 'contact' print which will be a positive by dyeline duplication. Sensitised paper is placed against the back of the original and whilst the two are fed by rollers through the machine they are exposed to a strong actinic light. The sensitised paper is then 'developed' by passing through an ammonia solution, and the linework will then appear as reasonably dense lines on a pale background, generally with a greyish-blue tinge. The chief defect of this process is the lack of permanence of the print, which will tend to fade and discolour, as well as a certain loss of sharpness in line and detail work.

References

ARNBERGER, E. (1966) *Handbuch der thematischen Kartographie*. Vienna.

BACHI, R. (1968) *Graphical Rational Patterns*. Jerusalem.

BAGROW, L. (1964) *History of Cartography*. London.

BENNETT, H. A. *et al.* (1967) The land map and air chart at 1/250,000. *Cart. J.* 4, 89–95.

BIRCH, T. W. (1964) *Maps, Topographical and Statistical*. Oxford.

BLALOCK, H. M. (1960) *Social Statistics*. New York.

BOARD, C. (1967) Maps as Models in *Models in Geography*, edited by CHOR-LEY, R. J. and HAGGETT, P. London.

BOUGHTON, W. C. (1969) *Contour Plans by Computer*. Christchurch, N.Z.

CARMICHAEL, L. D. (1964) Experiments in relief portrayal. *Cart. J.* 6, 18–20.

CARMICHAEL, L. D. (1969) The relief of map making, *Cart. J.* 6 18–20.

CHORLEY, R. J. and HAGGETT P. (eds.) (1967) *Models in Geography*. London.

CHRISS, M. and HAYES G. R. (1964) *An Introduction to Charts and their Use*. Glasgow.

CLARE, W. G. (1964) Map reproduction. *Cart. J.* 1, 42–8.

CLARK, D. (1963) *Plane and Geodetic Surveying*, Vol. I, 5th edition.

COLE, J. P. and KING, C. A. M. (1968) *Quantitative Geography*. London.

COLEMAN, A. (1965) Some technical and economic limitations of carto-graphic colour representation on Land Use maps. *Cart. J.* 2, 90–4.

CRONE, G. R. (1966) *Maps and their Makers*. London.

DAVIDSON, Viscount (1938) *Final Report of the Departmental Committee on the Ordnance Survey*. HMSO, London.

DICKINSON, G. C. (1969) *Maps and Air Photographs*. London.

DIELLO, J. *et al.* (1969) The development of an automated cartographic system. *Cart. J.* 6, 9–17.

DIXON, O. M. (1967) The selection of towns and other features on atlas maps of Nigeria. *Cart. J.* 4, 16–23.

DURY, G. H. (1960) *Map Interpretation*. London.

ENVIRONMENT, DEPARTMENT OF (1970) *Atlas of Planning Maps*, 2 volumes.

FROLOV, Y. S. and MALING, D. H. (1969) The accuracy of area measure-ment by point counting techniques. *Cart. J.* 6, 21–35.

GAITS, G. M. (1969) Thematic mapping by computer. *Cart. J.* 6, 50–68.

GALE, D. W. (1965) Register control in map production. *Cart. J.* 2, 68–74.

GEOGRAPHICAL MAGAZINE (October 1969) *Cartography for the 1970s*.

GUEST, A. (1969) *Advanced Practical Geography*. London.

HÄGERSTRAND, T. (1967) The computer and the geographer. *Trans. Inst. Brit. Geogr.* 42, 1–19.

HAGGETT, P. (1965) *Locational Analysis in Human Geography*. London.

HAGGETT, P. and CHORLEY, R. J. (1969) *Network Analysis in Geography*. London.

HARVEY, D. (1969) *Explanation in Geography*. London.

HINKS, A. R. (1947, 5th edition) *Maps and Survey*. Cambridge.

HODGKISS, A. G. (1970) *Maps for Books and Theses*. Newton Abbot.

HOWARD, S. (1968) A cartographic data bank for Ordnance Survey maps. *Cart. J.* 5, 48–53.

HUNT, A. J. (1968) Problems of population mapping. *Trans. Inst. Brit. Geogr.* 43, 1–8.

HUTCHINGS, E. A. D. (1970) *A Survey of Printing Processes*. London.

IMHOF, E. (1957) Die Vertikalabstände der Höhenkurven (in *Festschrift C. F. Baeschlin*, 77. Zürich).

IMHOF, E. (1965) *Kartographische Geländedarstellung*. Berlin.

IMHOF, E. (1966) *Schweizerischer Mittelschulatlas*. Zürich.

JOHANSSON, O. (1969) Further developments of producing photomaps as a basis for the economic map of Sweden. *Cart. J.* 6, 103–7.

JOLY, F. (1964) *Proposal for standardization of symbols on thematic maps*. (Paper presented to International Cartographic Association meeting, Edinburgh.) Paris.

KEATES, J. S. (1958) The use of type in cartography. *Surveying and Mapping* 18, 75–6.

KEATES, J. S. (1962) The perception of colour in cartography. *Proc. Edinburgh Cartographic Symposium*. Glasgow.

KERN, R. and RUSHTON, G. (1969) Mapit: a computer program for production of flow maps, dot maps and graduated symbol maps. *Cart. J.* 6, 131–7.

KILFORD, W. K. (1963) *Elementary Air Survey*. London.

LOCK, C. B. (1969) *Modern Maps and Atlases*. London.

LYONS, H. G. (1914) Relief in cartography. *Geogr. J.* 43, 233–48 and 395–407.

MACKAY, J. R. (1953) The alternative choice in isopleth interpolation. *Prof. Geogr.* 5, 2–4.

MAGEE, G. A. (1968) The Admiralty chart; trends in content and design. *Cart. J.* 5, 28–33.

MALING, D. H. (1968) How long is a piece of string? *Cart. J.* 5, 147–56.

MCGRATH, G. (1965 and 1966) The representation of vegetation on topographic maps. *Cart. J.* 2, 87–9 and 3, 74–8.

MCKAY, C. J. (1967) The pattern map. *Cart. J.* 4, 114.

MEINE, K. H. (1966) Aviation cartography. *Cart. J.* 3, 31–40.

MERRIAM, M. (1965) The conversion of aerial photography to symbolised maps. *Cart. J.* 2, 9–14.

MINISTRY OF HOUSING & LOCAL GOVERNMENT (in progress since 1955). *The Desk Planning Atlas of England and Wales*. London.

MONKHOUSE, F. J. and WILKINSON, H. R. (1971) *Maps and Diagrams*, 3rd edition. London.

ORDNANCE SURVEY (1951) *An Introduction to the Projection for Ordnance Survey Maps and the National Reference System*. London.

ORDNANCE SURVEY (1968) *Parcel Numbers and Areas on 1/2500 Scale Plans*. London.

ORDNANCE SURVEY (1969) *Its History, Organisation and Work*. Southampton.

ORDNANCE SURVEY, Map Catalogue, latest date, published annually.

OVINGTON, J. J. (1962) An outline of map reproduction. *Cart*. 4, 150–5.

PANNAKOEK, A. J. (1968) Problems encountered in producing the National Atlas of the Netherlands. *Cart. J*. 5, 8–15.

PETRIE, G. (1962) The automatic plotter and its consequences. *Proc. of the Edinburgh Cartographic Symposium*. Glasgow.

PETRIE, G. (1966) Numerically controlled methods of automatic plotting and draughting. *Cart. J*. 3, 60–73.

PIKET, G. (1968) The Water Control Map of the Netherlands. *Cart. J*. 5, 16–27.

RAISZ, E. (1962) *Principles of Cartography*. New York.

ROBERTSON, J. C. (1967) The Symap Programme for computer mapping. *Cart. J*. 4, 108–13.

ROBINSON, A. H. (1965) The future of the international map. *Cart. J*. 2, 23–6.

ROBINSON, A. H. and SALE, R. D. (1969) *Elements of Cartography*. New York.

ROSING, K. E. (1969) Computer graphics. *Area* 1, 2–8.

ROSING, K. E. and WOOD, P. A. (1971) *Character of a Conurbation*. London.

ROYAL SOCIETY (1966) *Glossary of Technical Terms in Cartography*. London.

SCOTT, L. (1969) Early experience of the photomapping technique. *Cart. J*. 6, 108–11.

STEERS, J. A. (1965) *An Introduction to the Study of Map Projections*.

STORRIE, M. and JACKSON, C. I. (1967) A comparison of some methods of mapping census data of the British Isles. *Cart. J*. 4, 38–43.

TARRANT, J. R. (1970) Comments on the use of trend surface analysis in the study of erosion surfaces. *Trans. Inst. Brit. Geogr*. 51, 221–2.

THORNES, J. B. and JONES, D. K. C. (1969) Regional and local components in the physiography of the Sussex Weald. *Area* 2, 13–21.

TOBLER, W. R. (1965) Automation in the preparation of thematic maps. *Cart. J*. 2, 32–8.

TOPFER, F. and PILLEWIZER, W. (1966) The principles of selection, a means of Cartographic generalisation. *Cart. J*. 3, 10–16.

WOOD, M. (1968) Visual perception and map design. *Cart. J*. 5, 54–64.

YOELI, P. (1967) Mechanisation in analytical hill-shading. *Cart. J*. 4, 82–8.

ZUYLEN, L. VAN. (1969) Production of photomaps. *Cart. J*. 6, 92–102.

Appendix I

The National Grid is printed on all Ordnance Survey Maps of Great Britain and is a system by which metric co-ordinates can be given to any point in these countries (see chapter 1, and especially fig. 1.5). Depending on the map scale, references can be read to the nearest metre, ten metres, kilometre, etc., and this is most easily done with the aid of a 'romer' and by measuring to the nearest grid lines west and south of the position under consideration, as follows:

To give a full grid reference at point P on the upper diagram on the next page:

1. Obtain from map margins the letters designating the 500 km and 100 km grid squares within which P lies.
2. Take the west edge of grid square in which P lies and read the figures printed opposite this line on north or south margin. Estimate tenths eastward.
3. Take the south edge of grid square in which P lies and read the figures printed opposite this line on east or west margin. Estimate tenths northward.

Examples of grid references on 1″ to 1 mile map, 1/250,000 (approximately ¼″ to 1 mile) map, 1/1,250 plan and 6″ to 1 mile map are given in this appendix. Note that these figures are not to scale but have been reduced to slightly more than half size.

The extracts given on the following pages show the full grid pattern on the four map scales, including the marginal subdivisions which can be used as scales for precise grid measurements. Other subdivisions of the various borders provide scales in different units, e.g. miles, chains, furlongs, yards, as well as providing information relating to latitude and longitude. These printed margins should be examined carefully to see what information they give; they are not purely decorative.

On pages 144 and 145 are reproduced the map legends or conventional sign cards for the large scale (1/1,250 and 1/2,500) plans of the Ordnance Survey. These contain all the necessary explanation of signs together with a 'romer' in the corner. Scale bars for distances in feet, chains and metres are also printed around the edges of the cards but these have not been reproduced here.

1″ TO 1 MILE MAP

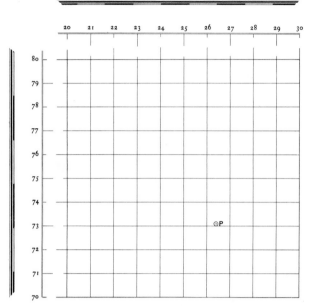

FULL GRID REFERENCE OF POINT P IS TO 264731

¼″ TO 1 MILE MAP

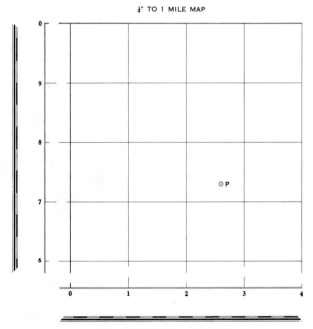

FULL GRID REFERENCE OF POINT P IS TQ 2673

1/1,250 SCALE PLAN

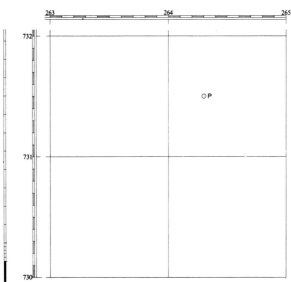

FULL GRID REFERENCE OF POINT P IS TQ 26437315

6″ TO 1 MILE MAP

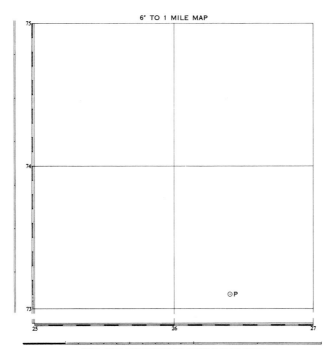

FULL GRID REFERENCE OF POINT P IS TQ 264731

REFERENCE CARD
NATIONAL PLANS AT SCALES OF
1:1250 or 50·688 inches to 1 Mile and 1:2500 or 25·344 inches to 1 Mile
(Information applicable to both scales except where otherwise stated)

SYMBOLS AND CONVENTIONAL SIGNS

.... Bracken *

.... Bush †

.... Coniferous Tree (Surveyed)

.... Coniferous Trees (Not Surveyed)

.... Coppice *

.... Deciduous Tree (Surveyed)

.... Deciduous Trees (Not Surveyed)

.... Furze †

.... Heath *

.... Marsh, Saltings

.... Orchard Tree

.... Osiers †

.... Osiers *

.... Reeds

.... Rough Grassland

.... Scrub *

.... Underwood †

Where space permits, tree and scrub symbols in woods are shown in groups of Four, Three, Two or One to indicate Close, Medium, Open or Scattered density of trees or scrub.

† Used on plans surveyed or revised prior to 22·4·63.

* Used on plans surveyed or revised since 22·4·63.

.... Antiquity (site of)

►►►► ..... Arrow showing direction of water flow

↑ B M .... Bench Mark (Normal)

↑ F B M .... Bench Mark (Fundamental)

⊗ .... Cave Entrance

▣ .... Electricity Pylon

· ts .... Permanent Traverse Station

· rp .... Revision Point (Instrumentally fixed)

↑ rp .... Revision Point and Bench Mark
(Coincident)

+ .... Surface Level

△ .... Triangulation Station

∫ .... Area Brace (1:2500 scale only)

Perimeter of built-up area with
single acreage (1:2500 scale only)

Levelling Information: Altitudes of bench marks and surface levels are given in feet above the Newlyn Datum. Bench Mark Lists, which may contain later levelling information, are obtainable from Director General, Ordnance Survey.

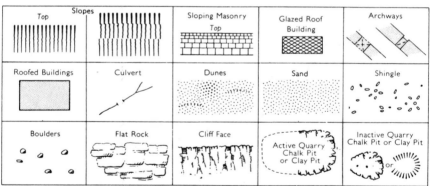

Slopes Top	Slopes	Sloping Masonry Top	Glazed Roof Building	Archways
Roofed Buildings	Culvert	Dunes	Sand	Shingle
Boulders	Flat Rock	Cliff Face	Active Quarry Chalk Pit or Clay Pit	Inactive Quarry Chalk Pit or Clay Pit

Walls, other than those of roofed buildings, which are less than 1 metre thick at 1:1250 scale or 1·4 metres thick at 1:2500 scale, are shown by a single line representing the centre of the wall.
Walls which are in excess of these widths are shown by two lines, each representing one face. For roofed buildings, the lines on the plan normally represent the outer face of the walls at ground level.

Thick Wall   Thin Walls   Represented by

for   Building

**See reverse for 1:2500 scales and Romer;
also for Boundary information and Abbreviations**

1:1250 Scale Romer: To Measure Ten Metre Reference
Place arrowhead on Point with edges of scale parallel to grid lines on plan. Move the scale S and W until divisions coincide with grid lines. Read scale at these points, horizontally first and vertically second.

9   8   7   6   5   4   3   2   1

BOUNDARY INFORMATION

England & Wales

▬ ▬ ▬	County Boundary (Geographical)
· ▬ · ▬ ·	County & Civil Parish Boundary coterminous
· ─┼─ · ─┼─ · L B Bdy	Administrative County or County Borough Boundary
○ ○ ○	London Borough Boundary
M B Bdy U D Bdy R D Bdy	County District Boundaries based on civil parish
· · R B Bdy · · ·	Rural Borough (Borough included in a Rural District)

England, Wales & Scotland

· · · · · ·	Civil Parish Boundary
Boro (or Burgh) Const Co Const	Parly & Ward Boundaries based on civil parish
Boro (or Burgh) Const & Ward Bdy Co Const Bdy	Parly & Ward Boundaries not based on civil parish

Scotland

(Not with parish)	(Coincident with parish)	
▬ ▬ ▬	· ▬ · ▬ ·	County Boundary (Geographical)
Co Cnl Bdy	Co Cnl Bdy	County Council Boundary
Co of City Bdy ─┼─	Co of City Bdy ─┼─	County of the City Boundary
Burgh Bdy	Burgh Bdy	Burgh Boundary
▬ Dist Bdy ▬ ▬	· · Dist Bdy · ·	District Council Boundary

Where the boundary of an Admin Co, Co Boro or Co of City is coincident
with that of a geographical county, the symbol for the latter is shown.

Examples of Boundary Mereings

Symbol marking point where boundary mereing changes

· · · Und · · ·		Undefined boundary
· · · Def · · ·		Original boundary feature destroyed or defaced
C B .... Centre of Bank	C O C S ...Centre of Old Course of Stream	F W .... Face of Wall
C C .... Centre of Canal, etc.	C C S .... Centre of Covered Stream	S R .... Side of River, etc.
C D .... Centre of Ditch, etc.	4ft R H .... 4 feet from Root of Hedge	T B .... Top of Bank
C R .... Centre of Road, etc.	E K .... Edge of Kerb	Tk H .... Track of Hedge
C·S .... Centre of Stream, etc.	F F .... Face of Fence	Tk S .... Track of Stream

ABBREVIATIONS IN FREQUENT USE

B H	Beer House	N T	National Trust
B P, B S	Boundary Post, Boundary Stone	N T L	Normal Tidal Limit
Cn, C	Capstan, Crane	P	Pillar, Pole or Post
Chy	Chimney	P C	Public Convenience
D Fn	Drinking Fountain	P C B	Police Call Box
El P	Electricity Pillar or Post	P T P	Police Telephone Pillar
E T L	Electricity Transmission Line	P O	Post Office
F A P	Fire Alarm Pillar	P H	Public House
F S	Flagstaff	Pp	Pump
F B	Foot Bridge	S B, S Br	Signal Box, Signal Bridge
G P	Guide Post	S P, S L	Signal Post, Signal Light
H	Hydrant or Hydraulic	Spr	Spring
L B	Letter Box	S, S D	Stone, Sundial
L C	Level Crossing	Tk	Tank or Track
L Twr	Lighting Tower	T C B	Telephone Call Box
L G	Loading Gauge	T C P	Telephone Call Post
Meml	Memorial	Tr	Trough
M P U	Mail Pick-up	Wr Pt, Wr T.	Water Point, Water Tap
M H	Manhole	W B	Weighbridge
M P	Mile Post or Mooring Post	W	Well
M S	Mile Stone	Wd Pp	Wind Pump

H or L W M M T .... High or Low Water Mark of Medium Tides (England & Wales)
H or L W M O S T .... High or Low Water Mark of Ordinary Spring Tides (Scotland)
M H or L W .... Mean High or Low Water (England and Wales) ⎫ On plans surveyed or
M H or L W S .... Mean High or Low Water Springs (Scotland) ⎭ revised after 12.11.64.

Made and Published by the Director General of the Ordnance Survey, Chessington, Surrey. 1967.

Romer for 1:2500 Scale

9 8 7 6 5 4 3 2

1
2
3
4
5
6
7
8
9

Appendix II

French 1/20,000 and German 1/50,000 map legends (see over). Those sections of the two legends referring to linear symbols are shown on the left, together with the German romer, as printed on the 1/50,000 map sheets. Opposite are reproduced the conventional signs for the two map series, together with the scale bars (1/50,000 printed vertically). Note that both scale bars include a scale for slope measurement with an example of their use. In the German case the slope may be measured. (*a*) in *grads*, (*b*) as a percentage or (*c*) as a ratio.

(*a*) Various road symbols on French topographic map series, from Autoroute (motorway) to sentier (footpath)

(*b*) Railways, and other communication line symbols from the same French series.

(*c*) Boundaries (Grenzen) and communications by rail and road on German topographic map series.

(*d*) Instructions for grid referencing and 'romer' printed on German map.

(*e*) and (*f*) Other conventional signs from the French and German topographic map sheets.

(*g*) and (*h*) scale bars from the French and German map sheets respectively.

(c)

Grenzen

▶━━━━━━━▶	*Staatsgrenze*
┥┝┥┝┥┝┥┝┥┝	*Landesgrenze*
─ · ─ · · ─ · ─ · · ─ · ─	*Regierungsbezirksgrenze*
─ ─ ─ ─ ─ ─	*Stadt- oder Landkreisgrenze*
─ ─ ─ ─ ─ ─ ─	*Truppenübungsplatzgrenze*
····························	*Naturschutzgebietsgrenze*

(a)

Autoroute (largeur réelle)	
Routes nationales:	
d'excellente viabilité	N. 16
de bonne viabilité	N. 19
de viabilité moyenne	N. 27
Chemins repris:	
de bonne viabilité	C. R. 107
de viabilité moyenne	C. R. 204
de viabilité médiocre	C. R. 310
Ch^ins empierrés:	
régulièrement entretenu	
irrégulièrement entretenu	
Chemin d'exploitation	
Laie forestière	
Sentier muletier, Ligne de coupe	
Sentier	

Verkehrsnetz

Bahnhof / *Haltepunkt*	*Vollspurige Bahn, mehrgleisig*
	Vollspurige Bahn, eingleisig
	Schmalspurige Bahn
	Zahnradbahn
	Straßen- und Wirtschaftsbahn
○─○─○─○─○	*Seil- und Schwebebahn*
═ ═ ═ ═ ═ *im Bau*	*Vollspurige Bahn im Bau*
═ ═	*Autobahn*
10	*Fernverkehrsstraße*
L I O	*Straße I A mit Baumreihen*
L I O 2	*Straße I B mit Kilometerstein*
	Unterhaltener Fahrweg II A
═══════════	*Unterhaltener Fahrweg II B*
	Feld- und Waldweg
·········─▸◂─·········	*Fußweg mit Steg*

(b)

Chemins de fer:	
à deux voies	
à une voie	
à voie étroite	
en tunnel	
à crémaillère	
en construction	
abandonné	
Voies de garage	
Tramway	
Chemin de fer transporteur, Plan incliné	
Câble transporteur. Téléphérique	
Câbles transporteurs d'énergie électrique	
Pipe-line	

(d)

Planzeiger

Zum Ablesen ist die waagerechte Teilung so an eine waagerechte Gitterlinie zu legen, daß die senkrechte Teilung den zu bezeichnenden Kartenpunkt berührt. Dann ist an der waagerechten Teilung bei der nächsten linken senkrechten Gitterlinie der „Rechts"- Wert, und an der senkrechten Teilung der „Hoch"- Wert abzulesen. Der Rechtswert ist stets zuerst zu nennen.
Die Punktangabe erfolgt in Metern. Nicht ablesbare Werte sind bis zur Angabe des vollen Meters durch Nullen zu ersetzen.

Beispiel: Punkt p liegt in Metern:

„Rechts" $^{35}_{*}$26000 + 1400 = ^{35}27400 = (kurz:) 27400
„Hoch" ^{53}96000 + 1100 = ^{53}97100 = (kurz:) 97100

* *Kennziffer des Meridianstreifens*

e)

Signaux géodésiques	△ ○ 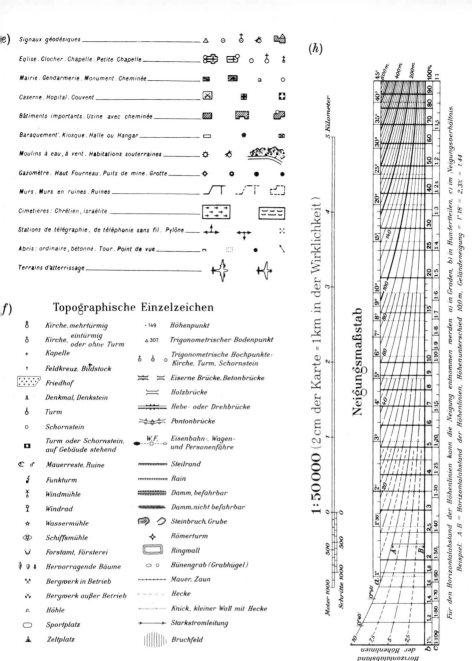
Eglise. Clocher. Chapelle. Petite Chapelle	
Mairie. Gendarmerie. Monument. Cheminée	
Caserne. Hôpital. Couvent	
Bâtiments importants. Usine avec cheminée	
Baraquement'. Kiosque. Halle ou Hangar	
Moulins à eau, à vent. Habitations souterraines	
Gazomètre. Haut Fourneau. Puits de mine. Grotte	
Murs. Murs en ruines. Ruines	
Cimetières : Chrétien, israélite	
Stations de télégraphie, de téléphonie sans fil. Pylône	
Abris: ordinaire, bétonné. Tour. Point de vue	
Terrains d'atterrissage	

(h)

Neigungsmaßstab

1:50000 (2cm der Karte = 1km in der Wirklichkeit)

Herausgegeben vom Landesvermessungsamt Rheinland-Pfalz 1962

Für den Horizontalabstand der Höhenlinien kann die Neigung entnommen werden a) in Graden, b) in Hunderteilen, c) im Neigungsverhältnis.
Beispiel: A-B = Horizontalabstand der Höhenlinien, Höhenunterschied 100 m, Geländeneigung = 1°18' = 2,3% = 1:44

f)

Topographische Einzelzeichen

	Kirche, mehrtürmig	• 149	Höhenpunkt
	Kirche, eintürmig oder ohne Turm	△ 307	Trigonometrischer Bodenpunkt
+	Kapelle		Trigonometrische Hochpunkte: Kirche, Turm, Schornstein
†	Feldkreuz, Bildstock		
	Friedhof		Eiserne Brücke, Betonbrücke
	Denkmal, Denkstein		Holzbrücke
	Turm		Hebe- oder Drehbrücke
○	Schornstein		Pontonbrücke
	Turm oder Schornstein, auf Gebäude stehend	W.F.	Eisenbahn-, Wagen- und Personenfähre
	Mauerreste, Ruine		Steilrand
	Funkturm		Rain
	Windmühle		Damm, befahrbar
	Windrad		Damm, nicht befahrbar
☆	Wassermühle		Steinbruch, Grube
	Schiffsmühle		Römerturm
	Forstamt, Försterei		Ringwall
	Hervorragende Bäume		Hünengrab (Grabhügel)
	Bergwerk in Betrieb		Mauer, Zaun
	Bergwerk außer Betrieb		Hecke
∩	Höhle		Knick, kleiner Wall mit Hecke
	Sportplatz		Starkstromleitung
▲	Zeltplatz		Bruchfeld

Nouvelle triangulation du Luxembourg.
Ellipsoïde international.
Projection Gauss-Kruger (système luxembourgeois).
Nivellement général du Luxembourg.
La partie allemande a été rédigée d'après
la carte d'Allemagne au 25.000°.

Echelle 1/20 000

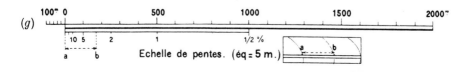

(g)

100ᵐ 0   500   1000   1500   2000ᵐ

10 5   2   1   1/2 %
a   b   Echelle de pentes. (éq = 5 m.)   a   b

Appendix III

Some characteristics of maps
United States of America

The topographic maps of the U.S.A. are produced by a branch of the Geological Survey – the Topographic Division which was founded in 1882 – and present both similarities with the maps of Western European countries described in chapter 6 and also some notable contrasts.

American map sheets are closely related to the graticule of the earth's lines of latitude and longitude with the actual sheet lines being based on a Conical projection (fig. App. III. 1). This arrangement permits the close inter-relationships between scales so that, for example, each map sheet at the Quarter Million (1/250,000) scale covers the same area as four sheets at the 1/125,000 scale. The basic map scale for any area depends to a great extent on the degree of development of the region. Thus for the state of Arizona the largest topographic map scale available for some areas is the Quarter Million series. These were prepared by the Army map service and give the following information –

Cultural features – towns, roads, railways, lettering in black. (Example of lettering styles fig. 3.6b.)
Relief features – contours, etc. – brown. Contour interval normally 200 feet
Water – blue

Built up areas are given a yellow fill and major woodland shown in green. With the exception of the yellow urban colouring these colour conventions are followed on the larger scale maps such as the 1/62,500 quadrangles which cover an area of 15′ latitude and longitude. Since the quarter million series are produced for 2° long × 1° lat quadrangles there are thus 32 of the larger scale maps to each of the smaller. The availability of maps at 1/62,500 scale is indicated on fig. App. III. 3 and it can be seen that the only areas not covered are the less accessible desert and mountain regions. The largest scale maps are for the $7\frac{1}{2}$′ quadrangles and there are thus four to each of the 1/62,500 sheets. This largest map scale is here at 1/24,000 although for some states of the U.S.A. the equivalent largest scale mapping is 2 inches to the mile (1/31,680).

The map sheets are usually described in terms of scale and sheet name and since cases arise of the same name applying to map sheets at different scales

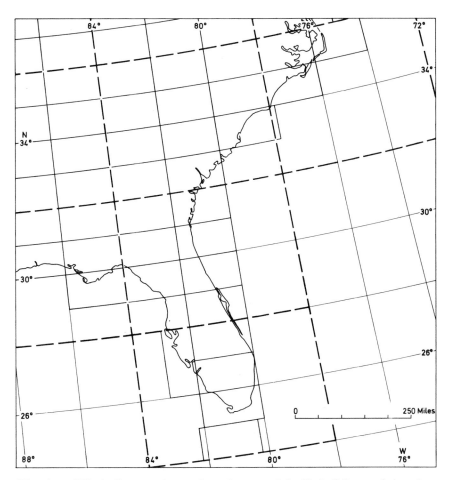

Fig. App. III. 1. Layout of map sheets for part of the United States of America. The heavy broken lines indicate the arrangement of the large one to one million map sheets and within these the lighter rectangles are the sheet lines of the quarter million map sheets (1/250,000 or approximately quarter inch to the mile scale).

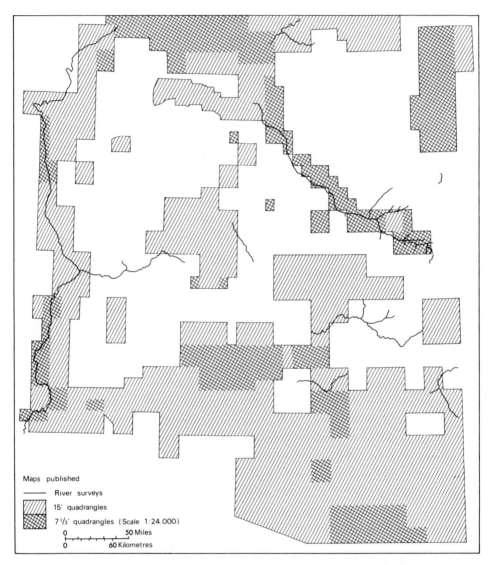

Maps published

—— River surveys

15′ quadrangles

7½′ quadrangles (Scale 1:24.000)

0   50 Miles

0   60 Kilometres

Fig. App. III. 2. Published maps for the state of Arizona at scales larger than 1/125,000. This map shows only the availability of maps at 1/62,500 (the 15′ quadrangles) and at 1/24,000 (the 7½′ quadrangles) as well as those sections of the rivers for which river surveys have been published.

it is necessary to take care in specifying these details – usually in terms of quadrangle size.

$7\frac{1}{2}'$ quadrangle/scale – usually 1/24,000 sometimes 1/31,680
15'   ,,   / ,, – 1/62,500
30'   ,,   / ,, – 1/125,000

Special sheets are produced for areas of importance, tourist interest, etc., and can be at larger scales (e.g. 1/6,000 for the Tombstone mining district). Plans and profiles of the major rivers are published at scales between 1/12,000 and 1/31,680 and some special sheets incorporating relief shading are also available.

The vertical contour interval on the larger scale topographic maps is frequently 20 feet but may be greater – depending on the area as well as the scale.

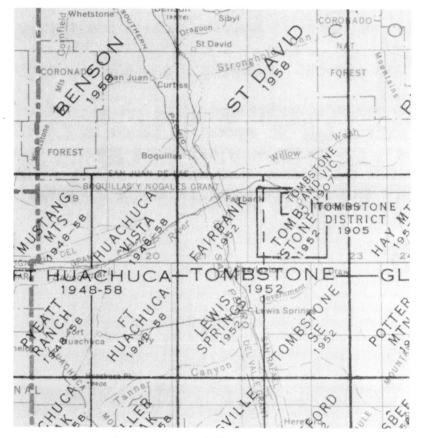

Fig. App. III. 3. Extract from the index map to the published topographic maps of the U.S. Geological Survey for the state of Arizona. This extract shows the 15' quadrangles, e.g. Huachuca, Tombstone, together with the $7\frac{1}{2}'$ quadrants into which they are divided. The dates of the editions are also shown and the background information includes some of the county and administrative divisions.

Appendix IV

Table of contents lists for (a) *Desk Planning Atlas of England and Wales and* (b) *Atlas of the Netherlands.*

(a) Atlas of Planning

Economic Planning Regions
Subdivisions of Economic
  Planning Regions
Land Classification
Types of Farming
Persons in Agricultural Occupations (1931)
Limestone
Population
  Changes by Migration 1921–1947
  Persistent Decrease 1921–1947
Total Population Changes 1921–1947
Population. Persistent Increase 1921–1947
Limestone Production
Population of Urban Areas 1951
Population. Volume of Change 1931–1951
Index to Administrative Areas in the
  1951 Census Sample Tables
Conservation of the Countryside
Regional Immigration
Coalfields 1952
Density of Agricultural Population 1931
Development Areas
Manufacturing Employment (1952)
Igneous and Metamorphic Rocks
Persons in Agricultural Occupations (1951)
Density of Agricultural Employment
  (1951)
Clothing
Chemicals and Allied Trades
Shipbuilding and Marine Engineering
Mechanical Engineering
Electrical Engineering and Electrical
  Goods
Cotton and Wool Textiles
Man-made Fibres and Other Textiles
Iron and Steel

Motor Vehicles and Cycles
Steel Sheets and Tinplate/Non-ferrous
  Metals
Metal Goods
Precision Instruments and Jewellery
Paper and Paper Products
Printing and Publishing
Food Industries
Drink and Tobacco
Pottery and Glassware
Furniture and Upholstery
Dwellings of Low Rateable Value in 1959
New Towns and Expanding Towns
Population Density 1951
Population. Volume of Change 1951–1961
Percentage Population Changes 1951–1961
Population of Urban Areas 1961
Rateable Value of Offices 1961
Rateable Value of Industry
Overcrowding
Dwellings of Low Rateable Value 1963
Housing. Average Household Size 1961
Institutions of Higher Education 1963–
  1964
Built-up Areas 1958
Population
  Rates of Increase 1931–1961
  Rates of Decrease 1931–1961
Housing
  Vacancy Rates 1961
  Ratio of all Dwellings to Households
  1961
Population. Total Change 1951–1961
Balance of Migration 1951–1961
Natural Change 1951–1961
Population in Private Households
Population. Age Structure Character of
  Local Authority Areas

Housing. Households in Shared Dwellings 1961

Estimated Gain or Loss of Retail Trade 1961

Status of Shopping Centres 1961

Population
Persons Aged 14 Years and Under
Persons Aged 65 Years and Over

Classification of Major Shopping Centres 1961

Housing. Household Arrangements 1961

Population. Socio-Economic Character 1961

Derelict Land 1964

Annual Migration of Employees June 1951–June 1964
Population. Migration 1960–1961
Duration of Residence at the 1961 Address

Housing
Overcrowding 1961
Occupied Dwellings Having 1–3 Rooms 1961

Gravel including Associated Sands

Workplace
Large Employment Centres 1961
Employment Centres with Net Gain of Workers 1961
Catchment Areas of Employment Centres

Civilian Population
Changes 1951–1965
Annual Changes 1961–1965
Total Change 1961–1965
Natural Change 1961–1965
Balance, mainly Migration 1961–1965

Housing
Households Owning their Accommodation 1961
Households Renting Accommodation from Local Authorities 1961
Households Renting Private Accommodation 1961

Principal Types of Housing Tenure 1961

(b) Atlas of the Netherlands

CARTOGRAPHY AND TOPOGRAPHY

1. The Netherlands and neighbouring countries
2. General map of The Netherlands
3. Fragments of different types of maps
4. Altimetric map
5. Administrative map
6. Fragments of historical maps

GEOLOGY, GEOPHYSICS, MINERAL RESOURCES

1. Geology of The Netherlands and adjacent regions
2. Tectonics of The Netherlands and adjacent regions
3. Geology
4. Tectonics (inset maps of South Limburg and eastern Gelderland)
5. Geological detail map
6. Mineral resources (inset maps of Limburg and eastern Gelderland)
7. Gravity and magnetic maps, including adjacent regions

GEOMORPHOLOGY

1. Geomorphology
2. Detailed maps of relief types

SOILS

1–11. Soil map on the scale of 1/200,000, in 11 sheets
12. General soil map, scale 1/600,000

CLIMATE

1. Precipitation
2. Precipitation, evaporation, and cloudiness
3–4. Temperature
5. Air pressure and wind

BIOGEOGRAPHY

1. Some detailed vegetation maps
2. Vegetation survey; Nature Reserves
3. Distribution of plants and animals

WATER MANAGEMENT

1. Drainage
2. Water District Boards

3. Water Management, water control
4. Geo-hydrology, ground-water movement
5. Water inlet, salinisation
6–7. Ground-water levels
8. Pollution of surface waters

HISTORICAL GEOGRAPHY

1–2. Archaeology and Prehistory
3. Habitability around A.D. 1300.
4. Expansion of the occupied area since A.D. 1300.
5. Changes in the courses of rivers
6. Changes in the areas and boundaries of the municipalities

SETTLEMENTS

1. Types and distribution of rural settlements; field patterns
2–3. Types of urban settlements

ANTHROPOLOGY, LANGUAGE, AND FOLK LIFE

1. Physical anthropology
2. Language, dialects, onomastics
3. Folk life

POPULATION, DEMOGRAPHY, SOCIAL AND ECONOMIC STRUCTURE

1. Population distribution
2. Population density
3–4. Population increase and decrease
5. Marital fertility
6. Age distribution
7. Public health
8. Housing
9. Religions
10. Political parties
11. Education
12–13. Economic activity of the population
14. Commuting
15. Personal income and property
16. Social work

AGRICULTURAL LAND UTILIZATION

1. Percentage arable land, grassland, horticultural land; agricultural areas
2. Distribution of forest and waste land
3. Types of farming, farm labourers, property, and leaseholding
4. Land consolidation, details of soils utilisation

5–6. Arable crops
7. Horticulture
8. Animal husbandry
9. Polder rates

PUBLIC UTILITIES

1. Drinking water
2. Gas
3. Electricity

FISHERIES

1. Location of fishing vessels and professional fishermen
2. Fishing grounds, types of fishing techniques, processing of fish

MANUFACTURING INDUSTRIES

1–2. Regional distribution of the economically active population engaged in manufacturing industries, mining, and construction
3–5. Regional distribution of manufacturing industries, mining, and construction
6–7. Location of manufacturing industries in towns

TRANSPORTATION AND COMMERCE

1. Distributional function of towns
2. Roads and canals
3. Density of motor vehicles
4. Road transport
5. Inland waterway transport
6. Goods transport
7. Seagoing shipping
8. Civil aviation
9. Passenger transport
10. Tourism and recreation
11. Isochrones
12. Foreign trade

PHYSICAL PLANNING

1. National plans (Western Metropolitan Area, Delta plan)
2. Regional plans (Limburg, North-West Overijssel)
3. Master plans (Amsterdam, The Hague, etc.)
4. Development plans and urban renewal plans (Rotterdam, Delft, etc.)
5. Plans for reclaimed land (Zuyder Sea, etc.)

Appendix V

The metric system: conversions factors and symbols

In common with several other textbook series *The Field of Geography* uses the metric units of measurement recommended for scientific journals by the Royal Society Conference of Editors.[1] For geography texts the most commonly used of these units are:

Physical quantity	Name of unit	Symbol for unit	Definition of non-basic units
length	metre	m	basic
area	square metre	m^2	basic
	hectare	ha	$10^4 \ m^2$
mass	kilogramme	kg	basic
	tonne	t	$10^3 \ kg$
volume	cubic metre	m^3	basic-derived
	litre	l	$10^{-3} \ m^3$, 1 dm^3
time	second	s	basic
force	newton	N	$kg \ m \ s^{-2}$
pressure	bar	bar	$10^5 \ Nm^{-2}$
energy	joule	J	kgm^2s^{-2}
power	watt	W	$kgm^2s^{-3} = Js^{-1}$
thermodynamic temperature	degree Kelvin	°K	basic
customary temperature, t	degree Celsius	°C	$t \ °C = T \ °K - 273.15$

Fractions and multiples

Fraction	Prefix	Symbol	Multiple	Prefix	Symbol
10^{-1}	deci	d	10	deka	da
10^{-2}	centi	c	10^2	hecto	h
10^{-3}	milli	m	10^3	kilo	k
10^{-6}	micro	μ	10^6	mega	M

The gramme (g) is used in association with numerical prefixes to avoid such absurdities as mkg for μg or kkg for Mg.

1. Royal Society Conference of Editors, *Metrication in Scientific Journals*, London, 1968.

Conversion of common British units to metric units

Length

1 mile = 1·609 km	1 fathom = 1·829 m
1 furlong = 0·201 km	1 yard = 0·914 m
1 chain = 20·117 m	1 foot = 0·305 m
	1 inch = 25·4 mm

Area

1 sq mile = 2·590 km^2	1 sq foot = 0·093 m^2
1 acre = 0·405 ha	1 sq inch = 645·16 mm^2

Mass

1 ton = 1·016 t	1 lb = 0·454 kg
1 cwt = 50·802 kg	1 oz = 28·350 g
1 stone = 6·350 kg	

Mass per unit length and per unit area

1 ton/mile = 0·631 t/km	1 ton/sq mile = 392·298 kg/km^2
1 lb/ft = 1·488 kg/m	1 cwt/acre = 125·535 kg/ha

Volume and capacity

1 cubic foot = 0·028 m^3	1 gallon = 4·546 l
1 cubic inch = 1638·71 mm^3	1 US gallon = 3·785 l
1 US barrel = 0·159 m^3	1 quart = 1·137 l
1 bushel = 0·036 m^3	1 pint = 0·568 l
	1 gill = 0·142 l

Velocity

1 m.p.h.   = 1·609 km/h
1 ft/s   = 0·305 m/s
1 UK knot = 1·853 km/h

Mass carried x distance

1 ton mile = 1·635 t km

Force

1 ton-force = 9·964 kN
1 lb-force = 4·448 N
1 poundal   = 0·138 N
1 dyn   = 10^{-5} N

Pressure

1 ton-force/ft^2 = 107·252 kN/m^2
1 lb-force/in^2 = 68·948 mbar
1 pdl/ft^2 = 1·488 N/m^2

Energy and power

1 therm	$= 105 \cdot 506$ MJ
1 hp	$= 745 \cdot 700$ W(J/s) $= 0 \cdot 746$ kW
1 hp/hour	$= 2 \cdot 685$ MJ
1 kWh	$= 3 \cdot 6$ MJ
1 Btu	$= 1 \cdot 055$ kJ
1 ft lb-force	$= 1 \cdot 356$ J
1 ft pdl	$= 0 \cdot 042$ J
1 cal	$= 4 \cdot 187$ J
1 erg	$= 10^{-7}$ J

Metric units have been used in the text wherever possible. British or other standard equivalents have been added in brackets in a few cases where metric units are still only used infrequently by English-speaking readers.

Index

In this index are included references to authors quoted in the text together with their cross-references in pages 137 to 139